THERAPEUTIC PROPERTIES OF
FERMENTED MILKS

ELSEVIER APPLIED FOOD SCIENCE SERIES

Biotechnology Applications in Beverage Production
C. CANTARELLI and G. LANZARINI

Progress in Sweeteners
T. H. GRENBY (Editor)

Food Refrigeration Processes: Analysis, Design and Simulation
A. C. CLELAND

Development and Application of Immunoassay for Food Analysis
J. H. R. RITTENBURG (Editor)

Microstructural Principles of Food Processing and Engineering
J. M. AGUILERA and D. W. STANLEY

Food Antioxidants
B. J. F. HUDSON (Editor)

Food Gels
P. HARRIS (Editor)

Forthcoming titles in this series:

Food Irradiation
S. THORNE

ORMATION SERVICES

THERAPEUTIC PROPERTIES OF FERMENTED MILKS

Edited by

R. K. ROBINSON

*Department of Food Science & Technology
University of Reading, Reading, UK*

ELSEVIER APPLIED SCIENCE
LONDON and NEW YORK

ELSEVIER SCIENCE PUBLISHERS LTD
Crown House, Linton Road, Barking, Essex IG11 8JU, England

Sole Distributor in the USA and Canada
ELSEVIER SCIENCE PUBLISHING CO., INC.
655 Avenue of the Americas, New York, NY 10010, USA

WITH 34 TABLES AND 5 ILLUSTRATIONS

© 1991 ELSEVIER SCIENCE PUBLISHERS LTD

British Library Cataloguing in Publication Data

Therapeutic properties of fermented milks.
 1. Food. Health aspects
 I. Robinson, R. K. (Richard Kenneth)
 613.2

ISBN 1-85166-552-8

Library of Congress Cataloging-in-Publication Data

Therapeutic properties of fermented milks/edited by R. K. Robinson.
    p. cm.—(Elsevier applied food science series)
  Includes bibliographical references (p.) and index.
  ISBN 1-85166-552-8
  1. Fermented milk—Therapeutic use. 2. Fermented milk—Microbiology. I. Robinson, R. K. (Richard Kenneth) II. Series.
RM234.T48   1991
615'.36—dc20

No responsibility is assumed by the publisher for any injury and/or damage to persons or property as a matter of products liability, negligence or otherwise, or from any use or operation of any methods, products, instructions or ideas contained in the material herein.

Special regulations for readers in the USA

The publication has been registered with the Copyright Clearance Center Inc. (CCC), Salem, Massachusetts. Information can be obtained from the CCC about conditions under which photocopies of parts of this publication may be made in the USA. All other copyright questions, including photocopying outside the USA, should be referred to the publisher.

All rights reserved. No part of this publication may be reproduced, stored in a retrieval system, or transmitted in any form or by any means, electronic, mechanical, photocopying, recording, or otherwise, without prior written permission of the publisher.

Photoset by Enset (Photosetting), Midsomer Norton, Bath, Avon
Printed in Great Britain at The University Press, Cambridge.

PREFACE

Nutrition, probably more than any other aspect of food science, thrives on a mixture of fact, speculation and myth, and the main reasons for this situation are not difficult to determine. Thus, nutritional studies on humans are notoriously difficult to perform, partly because of the problem of obtaining large test populations, and partly because any individual who does agree to participate will, to some degree, be a unique being. Certain basic metabolic pathways are obviously universal, but differences both qualitative and quantitative abound. The result of this diversity is that the same foodstuff may well affect different individuals in different ways, and perhaps in a manner that is too subtle to be recognised by standard clinical diagnoses.

It is this inherent variability between individuals that feeds the plethora of myths and 'old wives tales' that surround food habits. In reality, many of these 'much ridiculed' tales contain more than a 'grain of truth', particularly when assessed in a well-defined context. The early work of Metchnikoff, for example, suggested that the health and longevity of the Balkan peoples was founded on their regular consumption of yoghurt, while the literature of Russia contains similar accounts of illnesses avoided or cured by the inclusion of kefir or kumiss in the diet. Applied to different populations and/or products, such ideas may become untenable, and it is for this reason that, for the last fifty years or so, controversy has raged over the alleged health benefits of fermented milks.

Nevertheless, the basic hypothesis has never been disproved, and numerous authorities have claimed that fermented milks can be effective in combating a range of clinical or subclinical conditions. Indeed, evidence in favour of the view that certain 'beneficial' bacteria can have a health-promoting role has been steadily accumulating, and today few clinicians would feel disposed to state categorically that fermented milks cannot offer a health benefit over and above that indicated by a crude analysis for protein, vitamins or minerals. Obviously the type of bacterial culture is important, and clearly not every consumer will react in the same manner—even given the same pattern of consumption/complementary diet—but even so, genuine interest has replaced wholesale scepticism.

It is, therefore, a most opportune time to offer this critical examination of the dietary roles of fermented milks, and to this end, specialists from around the world have been invited to comment on the possible therapeutic properties of the major types found in the market-place. It must be admitted, of course, that definitive proof of health-promoting activity based on large-scale clinical trials is still lacking, and that the reaction of different test groups is often contradictory. Nevertheless, sufficient fragments of data have been accumulated to make the basic hypothesis look entirely plausible; so plausible in fact that it can no longer be simply dismissed.

It is a pleasure, therefore, to acknowledge with gratitude the efforts of the contributors that have made this book possible, and if it serves to keep alive the interest in the possible health-promoting properties of fermented milks, it will prove to have been a most worthwhile project.

R. K. ROBINSON

CONTENTS

Preface v

List of Contributors ix

1. Milk as a Food 1
   G. C. Cheeseman

2. Micro-organisms of Fermented Milks 23
   R. K. Robinson

3. The Therapeutic Effects of Various Cultures—An Overview . . . 45
   Livia Alm

4. Properties of Yoghurt 65
   Stanley E. Gilliland

5. Acidophilus Products 81
   Robert L. Sellars

6. The Health Potential of Products containing bifidobacteria 117
   J. A. Kurmann and J. Lj. Rašić

7. Products prepared with Lactic Acid Bacteria and Yeasts 159
   N. S. Koroleva

Index 181

LIST OF CONTRIBUTORS

L. ALM

   Medical Nutrition Department, Karolinska Institute, Stockholm, Sweden

G. C. CHEESEMAN

   Department of Food Science and Technology, University of Reading, Whiteknights, PO Box 226, Reading RG6 2AP, UK

S. E. GILLILAND

   Animal Science Department, Oklahoma State University, Stillwater, Oklahoma 74078-0425, USA

N. S. KOROLEVA

   International Dairy Federation, USSR National Committee for Dairying, 35 Lusinovskaya Street, M-93 Moscow 113093, USSR

J. A. KURMANN

   Agricultural Institute, Grangeneuve, 1725 Posieux, Switzerland

J. LJ. RAŠIĆ

   Rumenacka 106/I, 21000 Novi Sad, Yugoslavia

R. K. ROBINSON

   Department of Food Science and Technology, University of Reading, Whiteknights, PO Box 226, Reading RG6 2AP, UK

R. L. SELLARS

   Chr. Hansen's Laboratory Inc., 9015 West Maple Street, Milwaukee, Wisconsin 53214, USA

Chapter 1

MILK AS A FOOD

G. C. CHEESEMAN

Department of Food Science and Technology, University of Reading, UK

Milk is a complete food for newborn mammals. It is the sole food during the early stage of rapid development and as such contains those constituents, particularly proteins, fats and minerals, required for mammalian growth and development. As well as its nutritional properties, milk also affords protection against infection in the neonate by virtue of the protective factors it contains.

For young humans after weaning, milk, especially cows' milk, still plays an important dietary role providing fat for energy, proteins for muscle development and minerals for bone development. As part of a mixed diet milk has a proven record as a healthy, nutritious food. For the adult, products derived from milk become more important as constituents of a nutritionally balanced diet.

The history of milk as a food for man most likely extends back to prehistoric times. Cave drawings estimated to be 15 000 years old show that cattle were known in prehistoric times and there is evidence that domestication of cattle occurred before 4000 BC. Illustrations found on artifacts discovered in the Middle East, in Egypt and Mesopotamia, show that different types of cattle and various aspects of dairy farming existed about 3000 BC.

Dairy farming became well established in Ancient Egypt, as records have been found listing a cattle census in 2800 BC and paintings, dated by egyptologists as being done about 2500 BC, show cows being milked.

In Europe, domestication and the development of dairy farming occurred much later. Cattle were kept on common land, and a number were culled in the winter because of the shortage of grazing and the need for meat. Milk was consumed warm directly after milking. Surplus milk was converted into products, such as butter and cheese, as a means of preservation and sold locally. It was not until the 18th and 19th centuries that development of a more widespread distribution of milk as a food for the population occurred.

As the population of towns increased and the Industrial Revolution spread, so the need to supply the urban population with milk increased. It was not, however, until the beginning of the 20th century with the acceptance and widespread development of commercial pasteurisation of milk, together with improvements in milk production practices, that milk became established as a wholesome and safe food.

CONSTITUENTS OF MILK

Milk as a complete food for the newborn contains all the main nutrients required of a food. The amounts of the various constituents of milk are species dependent, and cows' milk differs in content from human milk in several important areas. For example, although the total fat content is similar, human milk fat contains a higher proportion of the long-chain unsaturated fatty acids. The protein fraction of human milk also contains a higher proportion of whey proteins and the lactose concentration is about twice that of cows' milk.

Cows' milk contains 3–4% protein, 3–6% fat, about 5% lactose and about 0.7% ash giving a total solids content of 11.5–15.5%. Many factors, such as breed characteristics, lactational, seasonal and dietary effects influence milk composition giving rise to variations in content.

In terms of dietary energy supply, a litre of milk of average composition, i.e. 3.5% protein, 3.8% fat and 4.7% lactose, would contribute about 660 kcal or 2750 kJ. Detailed nutritional composition data for milk and milk products is given by Paul and Southgate (1978) and Souci *et al.* (1981).

MILK PROTEIN

The importance of milk as a source of high quality protein for the growth and development of the young has already been mentioned. The nutritional importance is not only related to the amino acid composition of the proteins, but also to physico-chemical properties that enhance their value as a food. The proteins of cows' milk together with some of their properties are listed in Table I. The average protein content of milk is about 3.5%, but will vary slightly according to breed, stage of lactation, season of the year and the health of the animal.

TABLE I
PRINCIPAL PROTEINS IN COWS' MILK

Protein	Composition in skim-milk (g/litre)	Number of genetic variants	Molecular weight
α_{s1}-Casein	12–15	5	22 068–23 724
α_{s2}-Casein	3–4	4	25 230
β-Casein	9–11	7	23 944–24 092
κ-Casein	2–4	2	19 007 and 19 039
β-Lactoglobulin	2–4	8	18 205–18 363
α-Lactalbumin	0·6–1·7	2	14 147 and 14 175
Serum Albumin	0·4	1	66 267
Immunoglobulin G_1	0·3–0·6	—	153 000–163 000
Immunoglobulin G_2	0·05–0·1	—	146 000–154 000
Immunoglobulin A	0·05–0·15	—	385 000–417 000
Immunoglobulin M	0·05–0·1	—	1 000 000
Secretory Comp.	0·02–0·1	—	79 000

Source: Eigel et al., 1984.

The principal proteins in milk are the caseins, comprising about 85% of the total protein. They are a group of phosphoproteins which, by virtue of their physico-chemical properties, form stable complexes with calcium phosphate to produce protein mineral aggregates ranging in size from about 100 to 250 nm in diameter, the casein micelles. It is the micelles by reason of light scatter which give milk its white colour.

The phosphorylation of the caseins during synthesis is an essential biological process, as these groups are required to form stable casein micelles—necessary in the subsequent behaviour of the caseins as nutrients. The presence of phosphoserine residues provides a mechanism by which the mineral concentration of milk can be increased without causing instability, and also allows enzymes to preferentially act on the 'Ca-sensitive' caseins as principal substrates (West, 1986). Although nutritionally important, the caseins may have other specific functions. One suggestion is a relationship to the riboflavin-binding proteins.

A riboflavin-binding protein in chicken eggs, required for the transport of riboflavin across the yolk sac membrane, has eight serine phosphates arranged in sequence similar to those in casein. The protein is not involved with the vitamin binding site, but is presumed to be involved in binding to a recognition site on the membrane. No transfer of vitamins takes place if the protein is dephosphorylated. The similarity of the arrangement of the serine phosphates between the riboflavin-binding proteins and the caseins suggests the possibility that, although the caseins do not function as riboflavin binding proteins, they might act as a membrane recognition signal in the vitamin transfer process (West, 1986).

Fragments of β-casein have been isolated which have opiate-like activity in animal and isolated organ experiments. Some caseolytic strains of bacteria were found to produce these peptides, termed β-casomorphins, when incubated in cows' milk (Hamel et al., 1985). The principal member of the group β-casomorphin (1–7) has an amino acid sequence. TRY-PRO-PHE-PRO-GLY-PRO-ILE. However, no evidence has been found of the presence of β-casomorphins in human plasma after consuming cows' milk or milk products. Nevertheless, it has been speculated that these peptides may act as local regulators of intestinal function, in other words as 'food hormones' (Morley, 1982; Teschemacher et al., 1986).

Organisation of the micelle structure gives a highly concentrated arrangement of proteins, calcium and phosphate having an open, sponge-like structure which permits access of digestive enzymes. The k-casein linkage between residues 105 and 106 is uniquely susceptible to proteolytic attack. The hydrophilic portion of the molecule is released and the hydrophobic portion remains in the micelle. This change leads to micelle destabilisation and to the formation of a coagulum. The curd formed in the stomach of the young by the action of chymosin and pepsin is, however, less accessible to subsequent enzymic digestion than if the components were all soluble. The curd formation allows for greater retention in the stomach and a slower and prolonged digestion and release of nutrients (Poiffait & Adrian, 1986).

The curd formed from bovine milk is harder than that formed from human milk, and its digestion and release of nutrients is slower. Ca^{2+} absorption in the gut has been shown to be influenced by the presence of the casein phosphate groups, because if the casein is dephosphorylated, the absorption of Ca^{2+} is less. There is a high calcium content in bovine milk, about 32 mMoles, but little is present as the free ion Ca^{2+}. In the soluble phase it is complexed with citrate, and in the micellar phase as colloidal calcium phosphate.

The bioavailability of divalent metal ions such as Ca^{2+}, Zn^{2+}, Mn^{2+} and Fe^{2+} is influenced by their strong binding to the caseins. Although Zn^{2+} is released under acid conditions, it is retained in a chymosin formed coagulum.

Whey proteins, which are globular proteins, comprise the rest of the protein fraction. Most of the whey proteins, such as β-lactoglobulin and α-lactalbumin are synthesised in the mammary gland, whilst some of the rest, such as serum albumin and immunoglobulin, are derived from the blood.

Whey proteins constitute about 20% of the total milk protein, but contribute proportionally greater amounts of essential amino acids. There are 511 mg of essential amino acids/g of casein, but whey proteins contain 609 mg/g of protein. Whey proteins per gram contain 60% more of the sulphur-containing amino acids, methionine plus cysteine, than the caseins.

Although the principal whey proteins, α-lactalbumin and β-lactoglobulin, have an important nutritional role, other functions for these proteins have been postulated. It was shown over 20 years ago that α-lactalbumin could be substituted for one of the subunits of the enzyme lactose synthetase and enzyme functionality maintained, suggesting an enzymic role for this protein (Ebner et al., 1966).

More recently crystallographic analysis of baboon α-lactalbumin has shown that there is similarity with egg-white lysozyme. The amino acid sequences are similar, and it has been suggested that the class C lysozymes and α-lactalbumin are homologous proteins and that α-lactalbumin evolved from a lysozyme precursor (Smith et al., 1987).

The structure of the crystal form of β-lactoglobulin shows remarkable similarity to a retinol binding protein found in blood plasma. Such homology may suggest that β-lactoglobulin could have a role in vitamin A transfer (Papiz et al., 1986). However, any metabolic function of these two whey proteins remains as yet unproven.

Milk contains a number of protective proteins whose function is to protect against infection of the mammary gland and/or protect the neonate (Reiter, 1985). Lysozyme, a basic protein of about 15 000 daltons is present in cows' milk in very small quantities compared with human milk, where concentrations of about 40 mg/100 ml are found. This protein can attach to bacterial surfaces and, in the presence of anions, especially thiocyanate SCN^- and bicarbonate HCO_3^-, has lytic activity. Lactoperoxidase, a glycoprotein of about 77 500 daltons, occurs at about 3 mg/100 ml, and for its inhibitory activity requires the presence of peroxide, e.g. H_2O_2, and thiocyanate. Oxidation products of SCN^- are formed of which hypothiocyanate $OSCN^-$ is probably the most important. This anion attaches to the inner membrane of bacteria and inhibits membrane transfer of nutrients.

Lactoferrin, a glycoprotein of about 76 500 daltons, with two metal binding sites, each able to bind Fe^{3+}, occurs in cows' milk at concentrations between 2 and 35 mg/100 ml. It can only inhibit bacteria that have high iron requirements, such as the coliforms, and its iron binding capacity depends upon the presence of bicarbonate HCO_3^-. Citrate counteracts the bacteriostatic activity unless the concentration of bicarbonate is high.

Xanthine oxidase, a protein found associated with the milk fat globule membrane, may have some protective contribution. It is involved in the catabolism of purines and, in this reaction, superoxide (O_2^-) and H_2O_2 are generated. The synergistic action of the protective proteins with the immunoglobulins gives, overall, greater protection than does the sum of the individual components.

Immunoglobulin of the types IgA, IgG and IgM, together with free secretory component (FSC), are found in milk. Colostrum contains high concentrations of these proteins, while in milk, concentrations of about 14 mg of IgA, 60 mg of IgG_1, 2 mg of IgG_2, 5 mg of IgM and 5 mg of FSC per 100 ml of milk have been reported (Eigel *et al.*, 1984). The immunoglobulins enter the milk through the mammary secretory cells, but they are synthesised by plasma cells derived from lymphocytes. These plasma cells are located in various sites, but those synthesising IgA in particular appear to be situated in the mammary tissue.

A large adhesive glycoprotein, fibronectin, molecular size 233 000 daltons, has been isolated from milk where it occurs at a concentration of 2 μg/ml, and from colostrum where it is present at 30 μg/ml (Sato & Hayashi, 1985). It appears to be similar to blood plasma fibronectin which plays a role in the non-immune, self-defence system.

The protective factors being proteins are denatured by heat treatment, and under the conditions used in the preparation of a fermented milk, i.e. 85°C for 30 min or 95°C for 10 min, little could be expected to survive.

Genetic variants of all the principal proteins have been found. These protein variants usually differ in composition by substitution of one or more amino acids; up to nine for example in β-lactoglobulin B, or in the case of α_{s1}-casein A by the deletion of the 13 amino acids in positions 14 to 26 in the α_{s1}-casein molecule. The occurrence of some variants, e.g. α_{s1}-casein D is quite rare, and the effect of variants on the physical and nutritional properties of milk, particularly in bulk milk supplies, is virtually negligible (Eigel *et al.*, 1984).

Non-protein nitrogen constituents in milk comprise about 5% of the total nitrogen. The principal component of this fraction is urea, forming about half of the NPN. Other components include free amino acids, creatine and nitrate. Feed high in nitrogen will increase the levels of nitrate and urea in milk. The levels of urea in milk are thought to be important in the stability of heat-treated and concentrated milks; higher levels yielding more stable products, possibly through increased conversion of urea to cyanate and the subsequent reaction of cyanate with protein-bound lysine. The resultant change in charge, it is suggested, improves colloidal stability of the casein micelles (Muir & Sweetsur, 1978).

Milk proteins are relatively rich in essential amino acids. Individual protein fractions may vary and thus give rise to differences in amino acid content, but on average, milk protein contains about 46 g of essential amino acids per 100 g protein: 1·4 g tryptophan, 5·2 g phenylalanine 10·4 g leucine, 6·4 g isoleucine, 5·1 g threonine, 2·7 g methionine, 8·3 g lysine and 6·8 g valine. A seasonal variation has been reported in which the proportion of essential amino acids increases during the warmer months of the year, although total protein production during this period is less (Kirchmeier, 1973).

NUTRITIONAL VALUE OF MILK PROTEINS

There are a number of indices that can be used to compare nutritional properties of proteins. The most commonly used are: (a) biological value, where 100 g of dietary protein can replace the equivalent amount of adult body protein. A replacement value of 100 is given to whole egg protein, and higher or lower values reflect the quality of a protein; (b) net protein utilisation (NPU) determined from animal feeding experiments and is similar to biological value; and (c) protein efficiency ratio (PER) which relates to the weight gain produced by 1 g of dietary protein in a growing animal.

These indices give similar but not identical results, and a comparison of the values obtained for milk protein with some other food proteins is given in Table II.

In the case of the individual milk proteins, the PER value for casein is considered to be 2·9, whilst α-lactalbumin which is richer in essential amino acids has a value of 4 and β-lactoglobulin a value of 3·5.

Good quality protein, particularly animal protein, is especially important in the diet of the young and the elderly. Consumption of half a litre of milk a day will provide the recommended intake of essential amino acids for children and adolescents with the exception of methionine and cystine. In the case of these two sulphur-containing amino acids, milk

TABLE II
NUTRITIONAL VALUE INDICES

Protein	Biological value	NPU value	PER value
Egg	100	94	3·8
Milk	91	82	3·1
Beef	80	73	2·9
Soya	74	61	2·1
Wheat	54	41	1·5
Beans	49	39	1·4

provides 60% of the requirement for adolescents and 85% of the requirement for children. A daily intake of say 200–250 ml of milk in a mixed diet would ensure an adequate supply of essential amino acids for the young.

The requirement for protein increases in the elderly because the rate of protein metabolism is reduced, and there is a corresponding loss of muscle tissue. There is also an increased requirement for some essential amino acids, especially lysine and methionine, and these metabolic changes which take place after the age of 50 emphasise the need for a mainly animal protein diet. As there is also a reduction in energy requirement for the elderly, their diet should be balanced so that protein provides about 15% of the energy supplied. A daily intake of say 300 ml of milk and 50 g of cheese would, in a mixed diet, provide a significant part of the protein requirement; the use of low fat milk being advantageous.

MILK LIPIDS

Milk fat is secreted in the mammary gland as globules surrounded by a membrane derived from the secretory cell. The integrity of the membrane is effected by the cooling and processing of milk; part is released and also exchange takes place between the globule surface and other milk constituents. Fat globules have an average size of 2–6 μm, although there is a considerable distribution in size. The fat content of milk varies according to breed, season and to some degree in relation to the cows' diet (Faulkner et al., 1986). An average value of 3·8 g milk fat per 100 g of milk is a figure generally quoted. Milk fat consumption has tended to decrease in recent years partly due to health and diet implications, but also because of the relative cost with vegetable fats. Earlier quoted figures of milk fat contributing about 25% of dietary lipid, and milk and milk products contributing between 10–20% of total dietary energy (Renner, 1983) are likely to be higher than the present position, as suggested by the consumption figures for many developed countries.

The major lipid class found in milk fat is the triacylglycerides, which comprise some 97–98% of the total lipids. There are also small amounts of di- and mono-acylglycerols, free cholesterol and cholesterol esters, free fatty acids and phospholipids. The phospholipids are most commonly found associated with the fat globule membrane and other membraneous material occurring in milk. The main components are phosphatidyl-choline, phosphatidyl-ethanolamine and sphingomyelin with small amounts of phosphatidyl-serine, phosphatidyl-inositol and lysophospholipids also being present.

The fat in milk from ruminants contains relatively high concentrations of short-chain fatty acids—mainly butyric and hexanoic acids. These short-chain acids are rarely found in milk fat from non-ruminants. Expressed as a molar percentage, some 15% of milk fat is butyric acid and up to 40% of the triacylglycerol molecules have butyric acid as a constituent fatty acid.

The fatty acids of ruminant milk fat are obtained from two sources, the plasma lipids, which come from the diet and from stored lipid in adipose tissue, and *de novo* synthesis in the mammary gland. Ruminal fermentation gives rise to the low molecular weight precursors which are taken up from blood by the mammary gland. A major precursor, β-hydroxy butyric acid, is obtained by this route. The fatty acids synthesised in the mammary gland are principally acids of short-chain length, but synthesis includes some proportion of medium-chain acids up to C_{16}.

Medium-chain length fatty acids, such as myristic (C_{14}) and palmitic (C_{16}), also occur in significant amounts, but there are relatively low concentrations of the long-chain polyunsaturated fatty acids, such as linoleic (C18:2) and linolenic (C18:3).

The fatty acids found in cows' milk and human milk are compared in Table III.

The distribution of the fatty acids in the triacylglycerols of cows' milk is distinctive. The short-chain fatty acids are found mainly in the Sn-3 position. This arrangement is important in relation to the digestion of milk fat. The main products of the action of pancreative lipase on milk fat triacylglycerols are fatty acids from the Sn-1 and Sn-3 positions. Butyric and hexanoic acids released from the Sn-3 position are soluble in the

TABLE III
PRINCIPAL FATTY ACIDS IN COWS' MILK AND HUMAN MILK

	Fatty acid (Wt% of total)										
	$C4:0$	$C6:0$	$C8:0$	$C10:0$	$C12:0$	$C14:0$	$C16:0$	$C16:1$	$C18:0$	$C18:1$	$C18:2$	$C18:3$
Cow	3·3	1·6	1·3	3·0	3·1	9·5	26·0	2·3	14·6	29·8	2·4	8·0
Human	—	tr	tr	1·3	3·1	5·1	20·2	5·7	5·9	46·4	13·0	1·4

presence of calcium ions and will be rapidly absorbed into the body via the portal blood supply. The longer chain fatty acids from the Sn-1 position form relatively insoluble calcium salts and are less rapidly digested. In general, as the chain length increases, so the distribution favours firstly the Sn-2 position and then the Sn-1 position. For example, myristic (C_{14}) and palmitic acids (C_{16}) form a high proportion of the acids in the Sn-2 position, whilst stearic acid (C_{18}) is concentrated in the Sn-1 position.

Biohydrogenation of unsaturated fatty acids in the rumen of the cow means that the essential fatty acid content of milk fat is relatively low. On the basis that milk and milk products contribute, on average, about 35 g fat/day in the UK diet, the contribution of linolenic acid (C18:3) from dairy sources is about 0·5 g/day, or about 5% of the average total dietary intake of linolenic acid.

Fat-soluble vitamins, such as A, D, E and K, are found in milk fat, and β-carotene (a precursor of vitamin A) is better absorbed from milk fat than from many other foods. The β-carotene content of milk varies according to the feeding regime of the cow, being higher when the animal is grazing on pasture in spring and summer.

The unique fatty acid composition of milk fat is important in the palatability of milk and milk products. The textural property determining the milk's 'mouthfeel' relates to the size distribution of the milk fat globules together with the thermal or melting properties of the fat itself, which is dependent upon the fatty acid composition of the triacylglycerols. Flavour, perceived as odour and taste, is related to the more intense aromas of short-chain fatty acids such as butyric and caproic. The shorter chain acids being more volatile and water-soluble than long-chain acids.

MILK CARBOHYDRATES

Lactose, the principal carbohydrate in milk, is a disaccharide consisting of molecules of glucose and galactose. Cows' milk contains about 48 g of lactose per litre. Lactose supplies about 30% of the energy content of full cream milk, and as it is recommended that 55–60% of dietary energy be provided by carbohydrates; milk alone would not provide a suitable diet. Reducing the fat content, as in skimmed or semi-skimmed milk, increases the proportion of the energy supplied by the lactose and thus, in terms of energy supply, such products provide a better balanced food. Lactose has about one-third of the sweetening power of sucrose, but when hydrolysed to the constituent monosaccharides this increases to about two-thirds.

Small amounts of other carbohydrates occur in milk—the monosaccharides glucose and galactose are each found at concentrations of about 100 mg/litre. A number of oligosaccharides are also found whose total concentration is about 100 mg/litre. N-acetyl neuraminic acid occurs in the protein molecule of K-casein, and is found in the peptide fraction released by the enzyme chymosin from K-casein when the milk coagulation process is initiated.

Lactose is considered to play a role in calcium absorption. This is thought to be due, in part, to the effect of lactic acid, a major metabolic product of lactose, in bringing about a lower pH value in the intestine and thus increasing the solubility of the calcium salts. Its effect may also be due partly to the formation of soluble lactose–calcium complexes which could facilitate the transfer of calcium across the intestinal mucosa.

Incorporation of lactose in the diet in animal experiments has resulted in reducing symptoms of calcium deficiency, increasing blood calcium levels and reducing skeletal calcium losses. Extrapolation of animal experiments to human dietary metabolism is difficult, but it has been suggested that calcium in milk is better digested and utilised by humans than calcium in many other foods (Renner, 1983).

The breakdown of lactose by the enzyme lactase, secreted in the middle and lower sections of the small intestine, produces glucose and galactose which form substrates for the intestinal flora. These metabolic changes, together with those involving the production of lactic acid, encourage the development of an acid tolerant, benign intestinal flora.

Newborn infants develop lactase activity in response to the lactose present in the mother's milk. After weaning, the secretion of this β-galactosidase enzyme diminishes and more than 90% of human adults exhibit some degree of lactase deficiency. This reduction of enzyme production and the resultant lactose malabsorption is particularly marked within certain ethnic groups.

Amongst the Caucasian races of Central and Northern Europe, North America and Australasia, less than 10% are affected, whilst amongst Afro and Eurasian races the majority of adults are affected by lactose malabsorption. The nature of the diet may also have some influence, for example, Chinese born in Australia are less affected—some 50–60% suffer from lactose malabsorption, compared with their counterparts on mainland China where 95% of the adult population is affected.

When milk is heated, small amounts of lactose are converted into lactulose by isomerisation of the lactose molecule. Lactulose may occur free in solution or covalently bound to amino groups of the milk protein as ε-N-deoxy-lactulosyl-L-lysine. With mild heat treatments, such as pasteurisation, the amounts formed are negligible, but up to 70 mg/100 ml of lactulose may be found in UHT milk and up to 120 mg/100 ml in sterilised milks (Andrews, 1986). The reaction causes some reduction in the bioavailability of the milk lysine—mainly the whey protein lysine— however, as the lysine content is high, the nutritional loss is negligible except in extensively heat-treated milk.

MILK MINERALS AND TRACE ELEMENTS

Milk is an important source of essential nutrients, and in those countries where milk and milk products form a significant part of the diet, it makes a major contribution to dietary requirements of minerals. Some of the more important nutritionally are listed in Table IV. The values given are mean figures taken from analytical data reported in the literature.

Several of the elements in milk may occur in elevated concentrations because of contamination after milking, whilst others may be influenced by the cow's diet, the season and stage of lactation. Iron and copper levels can be increased through contact with metal containers and pipes, and

TABLE IV
MAJOR MINERALS AND TRACE ELEMENTS IN MILK

Component	Mean value (mg/100 g)	Range reported (mg/100 g)	Contribution to intake in UK diet (%)[a]
Sodium	45	12–90	7[b]
Potassium	150	37–200	22[b]
Calcium	120	58–250	60
Phosphorous	93	38–140	36
Magnesium	12·5	2–24	19
Iron	0·05	0·006–0·10	3
Zinc	0·36	0·08–0·70	22
Copper	0·01	0·001–0·07	6
Manganese	0·005	0·001–0·03	1
Selenium	0·002	0·0002–0·007	7
Iodine	0·0075	0·0005–0·04	36

[a] Milk and milk products.
[b] Milk.
Source: Hazell, 1985; Pennington et al., 1987; Bruhn & Franke, 1988.

iodine levels can be increased by contamination from iodophor disinfectants used as sanitation treatments of the udder and the milking plant. The iodine content is also influenced by its incorporation into feeds as a mineral supplement. Seasonal changes in diet, as well as mineral and trace element supplementation, affect the concentration of many of the elements, and the wide range of values reported in the literature reflect these influences (Pennington *et al.*, 1987; Bruhn & Franke, 1988).

Most of the trace elements in milk are found associated with the organic constituents. Some two-thirds of the iron is associated with the casein micelles, and a large proportion of the zinc is also found in this fraction; most of the remaining zinc is associated with the immunoglobulin fraction. Some zinc and iodine are also found free in the inorganic form.

Several trace elements, such as copper, zinc, manganese and iron, are also found associated with the fat globule membrane. Some of the iron occurs bound to a specific iron-binding protein, lactoferrin. Lactoferrin has a molecule size of 76 000 daltons and binds two iron atoms per molecule. The bioavailability of iron in cows' milk is five to seven times less than in human milk, and this difference is thought to be due to a much higher concentration of low molecular weight complexes of iron in human milk. A similar explanation is also suggested to account for the greater availability of zinc from human milk.

Contamination of feed with heavy metals, such as lead, cadmium and mercury, does not result in significant increased concentrations of these elements in the milk. Most of the lead eaten by the cow is excreted, and even on diets high in lead, values of only 50–100 µg per litre of milk are achievable. The cow acts as an effective filter of heavy metals and minimises their concentration in milk. Limits have been proposed for heavy metals in milk, these are: lead, 50 µg/litre; cadmium, 20 µg/litre; and mercury, 20 µg/litre. Only a small proportion, less than 10%, of heavy metal intake in human diet is attributable to milk and milk products (Renner, 1983).

NUTRITIONAL ROLE OF MILK MINERALS

In countries where milk and milk products form a significant part of the diet, some 60–90% of the dietary calcium requirement is obtained from these foods, 30–45% of the phosphorous and 20–25% of the magnesium. In the average UK diet, milk contributes about 7% of the sodium and 22% of the potassium obtained from foods before cooking (Hazell, 1985).

Because of the range of values found in milk for many trace elements, the contribution from milk and its products of these elements is less precise. However, probably, on average, about one-third of the iodine, a quarter of the zinc, cobalt, chromium and nickel, and smaller quantities (in the 2·5–10% range) of copper, fluorine, selenium, iron and manganese are contributed (Renner, 1983; Flynn & Power, 1985).

The absorption of milk calcium during digestion is very efficient; this is because much of the calcium is bound to the milk casein proteins and is made available gradually; this is probably the same for other organically-bound divalent metal ions, such as zinc, iron and manganese. Furthermore, other components in milk, such as lactose, vitamin D and citric acid, assist in the calcium absorption process. Milk in terms of nutrient density is the best readily-available source of calcium (Poiffait & Adrian, 1986). The importance of dietary calcium has been emphasised in relation to the development of osteoporosis, particularly in elderly women, and also in relation to the development of hypertension, high blood pressure and cardiovascular problems. Some reports recommend substantial increases in the recommended daily dose of calcium in the diet, from 800 mg/day to 1–2 g/day for susceptible sectors of the population, as a means of reducing the incidence and severity of these problems.

An inverse relationship has been shown between blood pressure and Ca^{2+} intake (Young et al., 1988). Low dietary Ca^{2+} intake and abnormal systemic Ca^{2+} handling are associated with hypertension. In hypertension, Ca^{2+} metabolism is found to be altered at both the whole animal and cellular levels. Oral supplementation of Ca^{2+} can lower blood pressure in human and experimental hypertension. The interrelationship between Ca^{2+} metabolism and blood pressure regulation is not as yet understood, but the importance of a readily bioavailable source of calcium in the diet is well substantiated.

MILK VITAMINS

Milk is a rich source of vitamins, particularly of the B groups and especially of riboflavin, vitamin B_2. All known vitamins occur in milk, and the principal vitamins found are given in Table V. With the exception of vitamin C, losses during processing treatments, such as pasteurisation, UHT sterilisation and manufacture of dried milk powders, are relatively small. However, in-bottle sterilisation and the sterilisation of evaporated milk may cause significant losses of thiamine, vitamin B_6, vitamin B_{12}, folic

TABLE V
PRINCIPAL VITAMINS IN COWS' MILK

Vitamin	Mean value (mg/100 g)	Range reported (mg/100 g)
A	37	10–90
Carotene	21	5–40
B_1 (thiamine)	42	20–80
B_2 (riboflavin)	172	80–260
B_6 (pyridoxine)	48	17–190
B_{12} (cobalmin)	0·45	0·2–0·7
Nicotinic acid	92	30–200
Folic acid	5·3	1–10
Pantothenic acid	36	26–49
C (ascorbic acid)	1800	500–3000
D (cholecalciferol)	0·08	0·01–0·2
E (tocopherol)	110	20–200
K	3	tr–17
Biotin	3·6	1–7

Source: Renner, 1983.

acid as well as vitamin C. Losses of vitamin C during processing, particularly in the less severe thermal treatments, are mainly caused by oxidative processes and not specifically by the higher temperature (Renner, 1986; Sieber, 1986).

The concentration of the fat-soluble vitamins will vary according to the fat content of the milk. In a milk containing 3·25% fat, a vitamin A content of 0·33 mg/litre was reported and this compares with a value of 0·23 mg/litre in 2% fat milk (Kikafunda & Olson, 1982).

The increased awareness, by a large proportion of consumers, of diet-related health hazards has resulted in a decreased consumption of butterfat, and an increasing trend towards the consumption of low fat milks. Such dietary trends can result, unless corrected by alternative supplementation, in a reduction in the intake of fat-soluble vitamins. Whereas in industrialised countries a decade or so ago, milk and milk products contributed 12–45% vitamin A, 5–20% vitamin D and about 10% vitamin E of the dietary vitamin intake, it is likely that these figures would now be significantly lower (Renner, 1983).

Of the water-soluble vitamins, it has been calculated that milk and milk products contribute 6–20% thiamine, 35–70% riboflavin, 10–20%

pyridoxine, 2–5% nicotinic acid, 20–30% pantothenic acid and 4–13% ascorbic acid of the dietary vitamin intake (Renner, 1983).

As milk is a valuable source of vitamins, it has an important role in the nutrition and growth of children and adolescents. The provision of milk at school is government policy in many developed countries, and much nutritional benefit has been obtained by these programmes which help to ensure that any deficiencies in diet are overcome.

EFFECT OF PROCESSING ON NUTRITIVE VALUE

Most of the milk consumed or converted into products is first heat processed. The purpose of thermal treatments is firstly to ensure that the milk is free from pathogens, and secondly to enhance the keeping quality by reducing the number of spoilage organisms that may be present in raw milk.

The principal heat treatment used is pasteurisation, which involves the heating of milk at 60°C for 30 min or more usually the equivalent high-temperature, short-time treatment of 72°C for 15 s. Such treatments eliminate or drastically reduce the number of pathogens present, and significantly reduce the number of spoilage organisms. To ensure sterility and to destroy the more heat-resistant bacterial spores, higher temperatures are required. In-bottle sterilisation involves heating the milk at 110–115°C for 20–30 min, whilst the more recent procedure of ultra-high temperature (UHT) processes, employing aseptic filling of the milk into sterile packages, uses temperatures of about 140° for 3–5 s.

All heat treatments bring about some chemical changes in the milk, the extent of which depends upon the severity of the heat treatment (Andrews, 1986). The changes are small with treatments such as pasteurisation and involve little loss in nutrient value (Renner, 1986; Sieber, 1986). However, the more severe treatments, such as in-bottle sterilisation and the sterilisation of evaporated milk, cause significant losses of some nutrients.

Such changes include denaturation of protein, interaction of reducing sugars, changes in solubility of salts, oxidation of lipids and off-flavour developments. Some of these reactions affect the organoleptic properties of the milk and its products, others influence the physical properties.

An important area of heat-induced chemical change is the effect upon milk proteins, particularly the whey protein fraction. The main proteins involved are the immunoglobulins, serum albumin, β-lactoglobulin and α-lactalbumin, in order of increasing heat stability. Less whey protein is denatured during milder heat treatments, and the UHT processes are less severe in this respect than the in-container sterilisation processes.

Pasteurisation treatments can cause about 15% denaturation of the whey protein, and more severe forms of heat treatment, such as in-bottle sterilisation, about 75% denaturation. The heat treatments used for a fermented milk, i.e. 85°C for 30 min or 95°C for 10 min, can cause 50–75% denaturation. During denaturation, whey proteins form aggregates primarily through disulphide linkages and also form aggregates with other milk proteins. β-Lactoglobulin was considered by Sawyer (1969) to complex with casein and more specifically with κ-casein. Snoeren and Van der Spek (1977) have isolated complex aggregates from UHT milk containing mixtures of β-lactoglobulin, α_{s2}- and κ-casein. α-Lactalbumin has also been shown to complex with β-lactoglobulin at temperatures below 90°C (Baer et al., 1976) and under conditions of UHT processing (Melo & Hansen, 1978).

Although the denaturation processes are complex, it is clear that, after heat treatment, a portion of the denatured whey protein is attached to the casein. The complex formation between the whey protein and casein interferes with the coagulating and gel-forming properties of the milk. In the more severely heat-treated milks, this will effect in-vivo curd formation, retention and digestion.

Reactions between proteins and sugars also occur during heat treatments. Such reactions include the Maillard reaction; a complex series of reactions initiated with a condensation reaction between the carbonyl group of lactose and the ε-amino group of lysine. The reactions involved include cross-linking between proteins to form insoluble aggregates, the development of brown pigments, the melanoidins, and production of several new compounds, such as hydroxymethylfurfural, furosine and pyridosine.

Losses of lysine are reported to be a maximum of about 2% in pasteurised milk, 4–6% in UHT processed milks and about 12% in sterilised milks (Renner, 1979). As the concentration of lysine in milk is relatively high, the loss caused by pasteurising milk is insignificant nutritionally.

Most of the liquid milk sold retail is pasteurised, and the shelf-life of good hygienic quality pasteurised milk held at 4°C is 16–21 days. Oxidative deterioration and the development of stale flavours, together with off-flavours from products of lipolysis and proteolysis, renders the milk less palatable (Allen & Joseph, 1985; Jeng et al., 1988). During storage, the ascorbic acid content may be reduced by oxidation, and some riboflavin may be lost through photooxidation.

MILK AS A POSSIBLE HEALTH HAZARD

Raw milk has long been known as a potential carrier of a number of diseases, some of which, e.g. tuberculosis, may originate from the cow, while others, such as typhoid fever and salmonellosis, may originate by contamination at or after milking. The general practice of pasteurisation together with improvements in dairy husbandry, milking procedures and the hygienic packaging, storage and distribution of milk in the last 30–40 years have greatly reduced the health risk from milk-borne diseases. Nevertheless, occasionally outbreaks of disease occur either from the consumption of contaminated raw milk, or from heat-treated milk that was contaminated post-pasteurisation or by faulty heat processing. The reported outbreaks of disease in recent years have been largely attributed to organisms such as *Listeria, Campylobacter, Yersinia* and *Salmonella* (Galbraith & Pusey, 1984; Diesch, 1985).

Attention in the last decade or so has also been given to the more general aspects of health and mortality in relation to diet. A number of reports have appeared suggesting that milk consumption is a factor in the increase, in the last 50 or 60 years, of atherosclerotic coronary disease. The seven-fold increase in per-capita consumption of milk in the UK over that period has paralleled a seven-fold increase in coronary mortality. Although no cause and effect can be substantiated, many hypotheses have been presented implicating a variety of bovine milk constituents in the development of atherosclerotic coronary disease. Of these, milk fat has probably been the most prominent, but most other constituents, such as milk proteins (Davies, 1984), fat globule membrane (Davies, 1984), lactose (Segall, 1984), xanthine oxidase (Oster, 1984), milk oestrogens (Seely, 1984) etc., have also been suggested.

The conflicting data and views on diet and health, and on the part played by individual dietary constituents, make it difficult to draw any firm conclusions on the long-term impact that this may have on milk as a food. At the present time, the recommendation by various authoritative bodies that consumption of saturated animal fats should be reduced has resulted in lower consumption of full cream milk, butter and high fat dairy products in most of the industrial countries.

On the other hand, the positive nutritional attributes, such as its high quality protein content, the bioavailability of milk calcium and other minerals and its vitamin contribution, are likely to continue to make milk an important food.

CONCLUSIONS

Milk, particularly that of the cow but also from sheep, goat and buffalo, has served mankind well as a major food source. The improvements in dairy farming and in the processing, handling and distribution of milk especially in the last 50 or 60 years has greatly improved the hygienic and compositional quality of milk, both in the liquid form and in its products.

There is, nevertheless considerable variation in the consumption of milk solids by people from different countries and cultures. Reported data for consumption in 1986 (given in Table VI) clearly demonstrate this fact, and indicate the relative contribution that milk and its products make to national diets.

TABLE VI
CONSUMPTION PER HEAD OF MILK FAT AND PROTEIN
(Kg) IN 1986

	1986	
	Milk fat	Milk protein
Iceland	19·0	13·3
Finland	19·0	11·5
Denmark	18·0	9·1
France	15·9	8·7
West Germany	13·7	6·9
Netherlands	12·1	9·2
UK	11·5	7·5
Canada	9·5	7·3
USA	8·5	7·2
Spain	5·9	5·3
Japan	2·3	2·2

Source: Anon., 1988.

Although a number of problems have beset the dairy industry in recent years, such as overproduction in some areas of the world and the effect of diet and health issues upon the consumption of certain milks and milk products, the overall trend worldwide has been for the consumption of milk solids to increase. This trend is partly due to population increases and improved standards of living (which has increased consumption in some regions, such as South America and Asia) counteracting the reduction in per-capita consumption in some developed countries in Europe and North America, but also partly due to the increased diversification in the manufacture and use of dairy products, particularly incorporation into other foods and the development of lower fat and fermented products. These changes are helping to ensure that milk and its products maintain their position as a major food for a large part of the world's population.

REFERENCES

Allen, J. & Joseph, G. (1985). *J. Dairy Res.*, **52**, 469–87.
Andrews, G. R. (1986). *J. Dairy Res.*, **53**, 665–80.
Anon. (1988). *Bull. Inter. Dairy Fed.*, **226**, 22.
Baer, A., Oroz, M. & Blanc, B. (1976). *J. Dairy Res.*, **43**, 419–32.
Bruhn, J. C. & Franke, A. A. (1988). *J. Dairy Sci.*, **71**, 917–24.
Davies, D. F. (1984). In *Health Hazards of Milk*, ed. D. L. J. Freed. Baillière Tindall, London, pp. 201–12.
Diesch, S. L. (1985). *Proc. U.S. Anim. Health Assoc.*, **89**, 336–45.
Ebner, K. E., Denton, W. L. & Bradbeck, Urs. (1966). *Biochem. Biophys. Res. Comms.*, **24**, 232–6.
Eigel, W. N., Butler, J. E., Ernstrom, C. A., Farrell, H. M., Harwalker, V. R., Jenness, R. & Mc. L. Whitney, R. (1984). *J. Dairy Sci.*, **67**, 1599–631.
Faulkner, A., Brechany, E. Y., Mabon, R. M. & Pollock, H. T. (1986). *J. Dairy Res.*, **53**, 223–7.
Flynn, A. & Power, P. (1985). In *Developments in Dairy Chemistry—3*, ed. P. F. Fox. Applied Science Publishers, London & New York, pp. 183–215.
Galbraith, N. S. & Pusey, J. J. (1984). In *Health Hazards of Milk*, ed. D. L. J. Freed. Baillière Tindall, London, pp. 27–59.
Hamel, U., Kielwein, G. & Teschemacher, H. (1985). *J. Dairy Res*, **52**, 139–48.
Hazell, T. (1985). *World Rev. of Nutr & Dietetics*, **46**, 1–123.
Jeng, W., Bassette, R. & Crang, R. E. (1988). *J. Dairy Sci.*, **71**, 2366–72.
Kikafunda, J. & Olson, J. P. (1982). *Can. Inst. Food Sci. Technol. J.*, **15**, 41–6.
Kirchmeier, O. (1973). *Milchwiss.*, **28**, 440–3.
Melo, T. S. & Hansen, A. P. (1978). *J. Dairy Res.*, **43**, 419–32.
Morley, J. E. (1982). *J. Amer. Med. Assoc.*, **237**, 2379–80.
Muir, D. D. & Sweetsur, A. W. M. (1978). *J. Dairy Res.*, **45**, 37–45.
Oster, K. A. (1984). In *Health Hazards of Milk*, ed. D. L. J. Freed. Baillière Tindall, London, pp. 239–53.
Papiz, M. Z., Sawyer, L., Eliopoulos, E. E., North, A. C. T., Findlay, J. B. L., Swaprasandorao, R., Jones, T. A., Newcomer, M. E. & Kraulis, P. J. (1986). *Nature*, **324**, 383–5.
Paul, A. A. & Southgate, D. A. T. (1978). *McCance and Widdowson's The Composition of Foods*. Elsevier/North-Holland Biomedical Press, New York, pp. 70–81.
Pennington, J. A. T., Wilson, D. B., Young, B. E., Johnson, R. D. & Vanderveen, J. E. (1987). *J. Amer. Dietetic Assoc.*, **87**, 1036–42.

Poiffait, A. & Adrian, J. (1986). *Indust. Alimen et Agricoles*, **103**, 335–41.
Reiter, B. (1985). *Bull. Inter. Dairy Fed.*, **191**, 1–35.
Renner, E. (1979). Nutritional and biochemical characteristics of UHT Milk. In *Proceedings: Int. Conf. on UHT processing and aseptic packaging* No. 27–29, Dept. of Food Sci., North Carolina State University, Raleigh, NC, USA.
Renner, E. (1983). *Milk & Dairy Products in Human Nutrition*. W-GmbH, Volkswirtschaft-licher, Munchen.
Renner, E. (1986). *Bull. Inter. Dairy Fed.*, **200**, 27–9.
Sato, T. N. & Hayashi, M. (1985). *J. Dairy Res.*, **55**, 507–11.
Sawyer, W. H. (1969). *J. Dairy Sci.*, **52**, 1347–55.
Seely, S. (1984). In *Health Hazards of Milk*, ed. D. L. J. Freed. Baillière Tindall, London, pp. 213–29.
Segall, J. J. (1984). In *Health Hazards of Milk*, ed. D. L. J. Freed. Baillière Tindall, London, pp. 229–39.
Sieber, R. (1986). *Bull. Inter. Dairy. Fed.*, **200**, 30–4.
Smith, S. G., Lewis, M., Achaffenburg, R., Fenna, R. E., Wilson, I. A., Sundaralingam, M., Stuart, D. I. & Phillips, D. C. (1987). *Biochem. J.*, **242**, 353–60.
Snoeren, T. H. M. & Van der Spek, C. A. (1977). *Neth. Milk & Dairy J.*, **31**, 352–5.
Souci, S. W., Fachmann, W. & Kraut, H. (1981). *Food Composition & Nutrition Tables, 1981/82*, Wissenschaftliche Verlagsgesellschaft mbH., Stuttgart.
Teschemacher, H., Umbach, M., Hamel, U., Praetorius, K., Ahnert-Hilger, G., Brantl, V., Lottspeich, F. & Henschen, A. (1986). *J. Dairy Res.*, **53**, 135–8.
West, D. W. (1986). *J. Dairy Res.*, **53**, 333–52.
Young, E. W., Bukoski, R. D. & McCarron, D. A. (1988). *Proc. Soc. Exp. Biol. Med.*, **187**, 123–41.

Chapter 2

MICRO-ORGANISMS OF FERMENTED MILKS

R. K. ROBINSON

Department of Food Science and Technology, University of Reading, UK

Although the history of fermented milks can be traced back for thousands of years, it is only since the turn of the century that serious attempts have been made to understand the associated microbiology. The result of this interest has been that many products are now manufactured with well-defined starter cultures, and that the fermentation processes can be subject to a considerable degree of control at plant level. Nevertheless, it has always to be borne in mind that the production of a fermented milk is still a natural biological process, and the organisms involved are subject to the same variability as any other life-form. Obviously, the selection pressures put upon a given species in a factory situation may be better defined than for species found in other environments, but even so, variation is still an active component of the system. The fact that many facets of the metabolism of the lactic acid bacteria are plasmid-controlled offers an especially clear route for change (McKay, 1985), and the manufacturers of starter cultures are all too aware of just how easily the characteristics of a culture can alter.

It is also evident that the physiology of these same micro-organisms can be considered, at least in practical terms, as falling into two distinct areas:

(i) metabolic pathways that are comparatively well-defined, and in many cases, common to a number of different species within the 'lactic acid' group (Kandler, 1983); and
(ii) special biochemical activities which may:
  (a) have been observed in very few species, or perhaps only strains of one species;
  (b) be poorly understood in terms of intermediate metabolites or enzymes; and

(c) lead to end products whose significance for either the bacterium concerned or the consumer of the fermented milk is subject to speculation (Chandon, 1989).

This latter group of activities will be discussed elsewhere in the book, and hence the aim of this introductory contribution is simply to establish some of the basic characteristics of those lactic acid bacteria, and other organisms, that are employed in the manufacture of yoghurt and similar products.

THE IMPORTANT MICRO-ORGANISMS

A 'fermented milk' has been defined by the International Dairy Federation as a 'milk product prepared from milk, skimmed or not, with specific cultures; the microflora is kept alive until sale to the consumer, and may not contain any pathogenic germ'. In the present context, the 'specific cultures' are predominantly composed of bacteria, and the important genera are *Lactobacillus, Streptococcus, Propionibacterium, Leuconostoc, Bifidobacterium, Pediococcus* and *Acetobacter*. Yeasts figure in a limited number of starter cultures but, in general, the group tend to be regarded more as agents of spoilage. The important species from within these genera, at least as far as fermented milks are concerned, are shown in Table I, together with certain basic characteristics that could be important either to the producer, e.g. behaviour at different temperatures, or the consumer, e.g. the isomer of lactic acid that is normally released (FAO/WHO, 1967).

However, although these comparisons are entirely relevant, further contrasts can be found, and hence it is pertinent to highlight, at generic level, the distinguishing features of the important groups.

Lactobacillus

The members of this genus exist as regularly-shaped, non-sporing rods, varying in length from 1·6 to 11·0 μm (Gilmour & Rowe, 1990). They are micro-aerophilic, and as shown in Table I, a number of species are thermophilic. In terms of their metabolism, the lactobacillae fall into three groups, namely:

(a) *Group I*, which are obligately homofermentative, and produce only lactic acid from hexose sugars via the Embden-Meyerhof pathway; pentoses are not fermented.

(b) *Group II*, which are facultatively heterofermentative, in that while many species metabolise hexoses to lactic acid alone, others can also form acetic acid, ethanol and formic acid under conditions of glucose starvation. Pentose sugars can be fermented to lactic and acetic acids through the induction of a phosphoketolase.

(c) *Group III*, which are obligately heterofermentative in that the members of this group ferment hexose sugars to lactic and acetic acids.

Amongst the lactobacillae regularly employed for the manufacture of fermented milks, it is of note that all the thermophilic species belong to Group I, and it is interesting to speculate whether the limited competition encountered at higher temperatures is reflected in the rather restricted spectrum of sugars being fermented. The advantage for the consumer, however, is that the resultant products have a clean 'lactic' taste without the 'harsh' notes that can sometimes accompany heterofermentative fermentations.

In the present context, the best known of the Group I thermophiles are *Lactobacillus delbrueckii* sub-sp. *bulgaricus*, often referred to simply as *Lactobacillus bulgaricus*, and *Lactobacillus acidophilus*, both of which are widely used in starter cultures by the dairy industry, and may also impart some rather interesting properties upon the products so produced.

Although *Lac. bulgaricus* is used during the production of a number of Swiss and Italian cheeses, its presence is central to the derivation of good quality yoghurt. Thus, not only does its synergistic interaction with *Streptococcus thermophilus* ensure that the desired fermentation takes place within an acceptable time period, but one facet of its metabolism is the release of acetaldehyde, and in sufficient quantities to give up to 40 ppm in the final product. This latter compound is an essential element in the typical 'taste' profile of yoghurt, and perceived flavour intensity is usually linked with the relative concentration of acetaldehyde (Robinson, 1988). The production of antibiotics by some strains of *Lac. bulgaricus* has been reported by some workers (Reddy & Shahani, 1971), as has the ability of other strains to resist bile salts and become implanted on the intestinal wall of human beings, but these two characteristics have been noted in only a few strains.

In contrast, *Lac. acidophilus* has long been regarded as having therapeutic properties, and indeed the sale of Acidophilus Milk is often promoted by health-food outlets. Irrespective of whether the product is categorised as Sweet Acidophilus Milk, i.e. inoculated with *Lac.*

TABLE I

THE BACTERIA AND YEASTS ASSOCIATED WITH THE MANUFACTURE OF FERMENTED MILKS, AND CERTAIN CHARACTERISTICS THAT COULD BE OF COMMERCIAL SIGNIFICANCE

Species	Growth at 10°C	Growth at 45°C	Form of lactic acid	Utilisation of Sucrose	Utilisation of Lactose	Utilisation of Galactose
Lactobacillus						
Lac. delbrueckii sub-sp. *delbrueckii*	−	+	D(−)	+	−	−
Lac. delbrueckii sub-sp. *lactis*	−	+	D(−)	+	+	±
Lac. delbrueckii sub-sp. *bulgaricus*	−	+	D(−)	−	+	−
Lac. acidophilus	−	+	DL	+	+	+
Lac. helveticus	−	+	DL	−	+	+
Lac. fermentum	−	+	DL	+	+	+
Lac. casei sub-sp. *casei*	+	−	L(+)	+	+	+
Lac. plantarum	+	−	DL	+	±	−
Lac. brevis	+	−	DL	±	±	+
Lac. kefir	+	−	DL	−	+	−
Streptococcus						
Str. lactis sub-sp. *lactis*	+	−	L(+)	±	+	nd
Str. lactis sub-sp. *cremoris*	+	−	L(+)	−	+	nd
Str. lactis biovar *diacetylactis*	+	−	L(+)	−	+	nd
Str. thermophilus	−	+	L(+)	+	+	nd
Leuconostoc						
Leu. mesenteroides sub-sp. *mesenteroides*	+	−	D(−)	±	+	+
Leu. mesenteroides sub-sp. *dextranicum*	+	−	D(−)	+	+	±
Leu. mesenteroides sub-sp. *cremoris*	+	−	D(−)	−	+	±
Leu. lactis	+	−	D(−)	+	+	+

Organism								
Bifidobacterium								
Bif. bifidum	−	−	L(+)	±	+	+	+	+
Bif. longum	−	−	L(+)	+	+	+	+	+
Bif. infantis	−	−	−	+	+	+	+	+
Bif. breve			L(+)	+	+	+	−	−
Acetobacter								
A. aceti			−	−	−	−	−	−
Pediococcus								
Ped. pentosaceus	±	−	DL	±	±	+	+	
Ped. acidilactici	−	+	DL	±	+	+	+	
Propionibacterium								
Pro. freudenreichii sub-sp. Shermanii			−	−	−	+	+	+	+
Yeasts	(37°C)							
Kluyveromyces marxianus sub-sp. marxianus	−	+	−	+	+	+	−	−
K. marxianus sub-sp. bulgaricus	−	+	−	+	+	+	−	−
Candida kefyr	−	+	−	+	+	+	−	−
Saccharomyces cerevisiae	−	−	−	+	−	−	−	−
Torulaspora delbrueckii	−	−	−					

\+ Positive reaction by most strains.
− Negative reaction by most strains.
± Reaction varies according to strain.
Source: Tamime (1989).

acidophilus but not fermented, or has been manufactured by the traditional route, the aim is to ensure that the retail material contains a minimum of 1×10^5 cfu g^{-1}. This level is often quoted as being the 'therapeutic minimum' (Nahaisi, 1986), and although not an excessive count in microbiological terms, it is not always easy to achieve with an organism as difficult to grow in milk as *Lac. acidophilus*. It is for this reason, at least in part, that a number of manufacturers are employing *Lac. acidophilus* in conjunction with another species with a similar reputation for health-promotion, *Bifidobacterium bifidum*. In many cases, the two species are grown separately before incorporation into a product, such as butter or ice-cream, so providing the retail item with a high 'loading' of the desirable microflora. A more specialised usage involves the co-inoculation of heat-treated milk with strains of the two species that can grow together with some degree of mutual stimulation. This latter approach offers the manufacturer the advantage of being able to handle *Lac. acidophilus* with rather more ease than is usually possible. At the same time, the heterofermentative capacity of *Bif. bifidum* gives the end product a rather more complex flavour than that associated with the usual range of *acidophilus* products. Obviously, care must be taken to control the level of acetic acid and other metabolites secreted by *Bif. bifidum*, but otherwise, the system has much to commend it (Tamime & Robinson, 1988). It is notable also that both species are sensitive to acid conditions, and hence the maintenance of high counts in the retail products needs special attention.

The remaining *Lactobacillus* spp. in Group I are mainly used in the production of high-scald cheeses, but *Lac. helveticus* has been associated with the production of kefir (Koroleva, 1988), while its sub-species, *Lac. helveticus* sub-sp. *jugurti* can be, as the name suggests, employed as a starter for yoghurt.

The mesophilic species, *Lac. casei*, is the most widely known member of Group II, and it is employed for the production of yakult. It has a number of sub-species, namely *casei, pseudoplantarum, rhamnosus* and *tolerans*. Whether or not some of these sub-species should be elevated to the status of species is a matter of some debate, but what is important is that strains of both *Lac. casei* (Alexander, 1971) and *Lac. casei* sub-sp. *rhamnosus* (Schuler & Schuler, 1983) have been identified as members of the normal, human intestinal flora.

The heterofermentative lactobacillae found in fermented milks, such as *Lac. brevis* and *Lac. kefir*, are mesophiles whose role in the relevant fermentations is somewhat secondary. Thus, in kefir, for example, these species together with other lactobacillae, are often present at comparatively low levels, and their role in the derivation of the end product is far from clear.

Streptococcus
The individual cells of members of this genus are small—less than 2·0 μm—and round, and they often occur in pairs or chains. Most members of the group are facultative anaerobes, and in the present context, it is the so-called 'lactic streptococci' which are of most relevance. The majority of the lactic streptococci are mesophiles, and although 30°C is usually quoted as the optimum temperature for growth, it is often recommended that starter cultures should be grown at 22–23°C. One important exception to this generalisation is *Str. thermophilus,* for in this case, the optimum temperature for growth usually lies between 37 and 45°C. It is, therefore, the ideal species for incorporation into starter cultures for yoghurt, for the complementary component of the starter, *Lac. bulgaricus*, is similarly thermophilic. The synergistic interaction between the two organisms is, as has already been mentioned, essential for successful production on a routine basis, but it is notable that not all strains of these two species are physiologically compatible. When synergy has been detected, then a clear basis for the mutual stimulation appears to exist, in that while *Str. thermophilus* benefits from the proteolytic activity of *Lac. bulgaricus*, the acid conditions generated by *Str. thermophilus* along with CO_2 and, perhaps, formic acid promote the development of *Lac. bulgaricus* (Tamime & Robinson, 1985). Whether or not other factors are involved remains to be determined, as does the reason(s) for the apparent lack of interaction observed with some pairings, and also the ease with which *Str. thermophilus* will grow in certain cultures accompanied by other lactic acid bacteria, e.g. *Lac. acidophilus*. Indeed it is not improbable that subtle degrees of association could be identified in many food fermentations, and certainly there is circumstantial evidence of inter-specific interactions between members of the mesophilic streptococci.

Three sub-species of *Str. lactis* are widely employed as starters for fermented milks, and these are *Str. lactis* sub-spp. *lactis, cremoris* and *diacetylactis*. In some texts, *Str. lactis* is now regarded as the true species with *cremoris* as a sub-species and *diacetylactis* as a biovar; the latter designation is employed because *Str. lactis* and *Str. lactis* biovar *diacetylactis* are similar except for the ability of the latter to produce diacetyl from citrate (Marshall, 1987). As this latter facility is plasmid-controlled (Kempler & McKay, 1981), and hence easily lost, it is regarded as inappropriate to treat *diacetylactis* as a sub-species. It has further been proposed that the lactic streptococci should be placed in a separate genus, *Lactococcus*, so that the nomenclature of the important organisms may well become *Lactococcus lactis, Lac. lactis* sub-sp. *cremoris* and *Lac. lactis* biovar *diacetylactis* (Gilmour & Rowe, 1990)—and this latter classification may well find universal favour in due course.

In practice, both *Str. lactis* sub-spp. *lactis* and *cremoris* are employed to provide the desired level of lactic acid in a product, while the biovar *diacetylactis* is introduced as a source of aroma/flavour compounds. The derivation of diacetyl from citrate is the most important pathway involved, but some strains also release acetic acid as the result of citrate metabolism. At the low levels recorded, the acetic acid is a welcome addition to the flavour profile, and at the same time, its presence may increase the shelf-life of an end product by suppressing the growth of Gram-negative bacteria.

Although both sub-species of *Str. lactis* are essentially homofermentative, other by-products of their metabolism could be important in relation to the quality of fermented milks. The secretion of nisin, for example, by *Str. lactis* sub-sp. *lactis* may be active in reducing spoilage by Gram-positive organisms, such as *Bacillus* spp. (Hurst, 1983), while Babel (1977) suggested that diplococcin originating from strains of *Str. lactis* sub-sp. *cremoris* could be instrumental in inhibiting certain strains of *Staphylococcus aureus*. In a different direction, the quality of a number of Scandinavian fermented milks depends on the ability of certain lactic streptococci, notably *Str. lactis* sub-sp. *cremoris* to produce protein/carbohydrate polymers which, in turn, considerably enhance the organoleptic properties of the various retail products (Macura & Townsley, 1984; Tamime & Robinson, 1988).

It is also relevant that the mesophilic streptococci are highly susceptible to invasion by bacteriophage (Cogan & Accolas, 1990), and the phage relationships of the various species and/or strains of *Streptococcus* are critical to the selection of starter cultures for a given factory. In cheese factories, the use of phage-resistant strains and/or the introduction of strict rotations of strains is essential to ensure a minimum of vat failures (Heap & Lawrence, 1988), and it should be noted that the same species are employed in many fermented milks.

Leuconostoc
Members of this genus occur as Gram-positive cocci, usually associated

into pairs or chains. Glucose is fermented with the liberation of lactic acid, ethanol and carbon dioxide, and hence the presence of *Leuconostoc* spp. in dairy products is usually linked with flavour production rather than acidification. According to Tamime (1990), the most important species is *Leu. mesenteroides*, which has been divided into three sub-species, namely *mesenteroides, dextranicum* and *cremoris*. The latter sub-species is most frequently employed in dairy products, and it is used, along with other mesophilic starter organisms, in the manufacture of cottage and cream cheeses, cultured cream and buttermilk. Its selection for these roles depends on the facts that:

(i) under the acid conditions produced by the co-inoculated starter bacteria, i.e. pH < 5·0, *Leu. mesenteroides* sub-sp. *cremoris* releases diacetyl into the medium;
(ii) the other sub-species of *Leu. mesenteroides* are capable of forming dextrans, and the 'slimy' mouthfeel imparted by these polysaccharides may be undesirable in some products; and
(iii) *Leu. mesenteroides* sub-sp. *cremoris* is able to generate acetic acid from the citrate present in milk, and the presence of this acid in, for example, cottage cheese can suppress the development of spoilage organisms, such as coliforms and pseudomonads.

The remaining sub-species have both been employed in the production of kefir, mainly to impart the characteristic taste and aroma, but also to improve the body of the product through the secretion of low levels of dextran. However, this usage is confined to the new generation of processes designed to avoid the irregular fermentations that can be encountered with the traditional inoculum of 'kefir grains'; *Leuconostoc* spp. are not normal constituents of the microflora of kefir.

Bifidobacterium
Bifidobacteria occur naturally in the intestinal tract of human infants and adults, and they appear as Gram-positive, non-motile, non-spore-forming rods of variable appearance (Rašić & Kurmann, 1983). Although strictly anaerobic, sensitivity to oxygen differs between strains of the various species, but the inhibition of growth under acid conditions, i.e. < pH 5·5, is widely found.

Although a number of species of bifidobacteria are recognised as being of interest in respect of their potential therapeutic properties, strain variability does pose something of a problem for the taxonomist. This difficulty of identification could be important for the consumer, because according to Rašić & Kurmann (1983), the species composition of the human varies with age. Thus, *Bifidobacterium infantis* appears to be the dominant species in breast-fed infants, although *Bif. bifidum* and *Bif. breve* may also occur. In adults, *Bif. longum* and *Bif. adolescentis* are commonly found, along with biotypes of *Bif. bifidum*. The reasons for, and significance of, these changes is a matter for speculation, but as will be discussed later, there is increasing evidence that an intestinal microflora dominated by bifidobacteria can be regarded as beneficial to good health.

In an effort to meet this desirable objective, an increasing number of fermented milks are appearing on the market with *Bif. bifidum* or *Bif. longum* included in the starter culture, and this requirement poses a number of problems for the manufacturer. Thus, as with *Lac. acidophilus*, a therapeutic minimum count of $10^5/10^6$ viable cells per millilitre of product is regarded as essential, and to maintain this level throughout the anticipated shelf-life of the retail item means that either:

(a) the acidity of the milk must be low enough to ensure that the cells will survive for at least two/three weeks prior to consumption; or
(b) the cell count at the end of incubation must be sufficiently high to allow for up to 90% mortality during storage, and yet still leave a figure above the desired minimum.

These alternative approaches have both found favour with different companies, but it is clearly imperative that the selected route does not damage the vitality of the cells. Thus, even after ingestion, the cells will still have to survive the inhibitory influence of the gastric juice, and also the bacteriostatic/bacteriocidal effect of bile acids. It is fortunate, of course, that readily digested foods like fermented milks have a relatively short transit time in the stomach, but clearly the effect of other inhibitory factors may not be so short-lived. The germicidal activity of other foods in the diet must also be borne in mind, and hence employing specially-fermented milks to modify the intestinal microflora requires considerable attention to detail.

Other Genera

The other genera that are sometimes associated with fermented milks, such as *Pediococcus* or *Propionibacterium*, are of rather minor importance in relation to overall production, and tend to be employed in just one or two specific products. *Pediococcus acidilacti*, for example, is employed as one of the starter organisms in Biokys, a cultured milk from Czechoslovakia, although it is not clear why this species has been specifically selected as the principal acidifying agent (IDF, 1988).

Similarly, *Acetobacter aceti* occupies a rather restricted niche in the dairy industry, being found only in starters for kefir or kumiss. In the latter product, Dellaglio (1988) suggests that it may be no more than an occasional member of the active microflora, but with kefir, the presence of *A. aceti* is regarded as central to maintenance of the integrity of the kefir grains. According to Koroleva (1988), the acidification activity of the kefir grains may be higher when both streptococci and acetic acid bacteria are present, while the latter group may also be instrumental in increasing the viscosity of the retail milk. However, the main role of this Gram-negative, obligately aerobic species is probably to provide, by the oxidation of ethanol to acetic acid, a more complex flavour profile.

The role of the yeasts detailed in Table I is again confined to kefir and kumiss, although it is of note that their activity in the two products is not necessarily identical. In both cases, however, it is the main role of the *Kluyveromyces* spp. to produce alcohol from the lactose, although some strains of *Candida kefyr* can also ferment lactose along with glucose and galactose. In kefir, of course, the *Candida* has a unique niche, in that it forms a symbiotic relationship with the various bacteria present in the kefir grains, and is probably the principal yeast associated with their structure (Robinson & Tamime, 1990). The remaining yeasts are seemingly little more than 'occasional' members of the microflora of these products, and certainly they appear to be poorly adapted to the substrate.

It is clear, therefore, that the range of starter bacteria employed for the production of fermented milks is, in reality, comparatively narrow and, furthermore, that the choice is restricted even more by the specific requirements of each product. Thus, the aims of employing starter cultures can be summarised as:

(1) preservation of the milk against putrefaction and, at least with products of pH 4·5 or less, suppression of pathogens;
(2) changing the flavour of the raw material and, perhaps, deriving a food of more widespread appeal to a given population;
(3) improving the nutritional value of the raw material either through the mechanism of processing, e.g. the heat treatment of the milk (Robinson, 1977), or through the action of the starter culture (Tamime & Deeth, 1980); and
(4) production of the fermented milk that can act as a vehicle by which a micro-organism with health-promoting properties can be introduced to a target group of consumers.

To achieve one or more of these aims, the starter culture must contain an active, viable population of the desired bacteria, so that given the correct selection of processing parameters, the bacteria will continue to grow and metabolise throughout the incubation period of the process milk. The survival of the organisms throughout the shelf-life of the retail product is equally important, and hence the maintenance and handling of starter cultures is a subject of especial concern.

Production of Starter Cultures

The isolation of any particular species or strain is often the result of a chance examination of a batch of product made by a traditional method. However, once the organism(s) has been brought into cultivation, the essentials are then to:

(a) preserve the specific organism in such a manner that its inherent physiological characteristics do not alter;
(b) ensure that it is available, if desired, for widespread usage throughout the dairy industry; and
(c) ensure that the means are available for the organism to be employed to manufacture an acceptable retail product.

Specialisation within the dairy industry has led, in recent years, to a division of these functions, so that while the latter is the concern of every factory producing fermented milks, the former have become the preserve of specialist suppliers of cultures. Some of the techniques employed in the isolation and handling of specific strains have been discussed elsewhere (Gasson, 1986; Heap & Lawrence, 1988; Tamime, 1990), and hence attention can be restricted to the handling of cultures in the factory.

Inoculation of the Process Milk

There are basically three systems available to introduce the desired culture into the fermentation vessel:

(i) the growth, on-site, of increasing volumes of culture culminating with a bulk starter, the volume of which will vary between 1 and 5% (v/v) of the final process milk; the figure selected will, in practice, depend on the nature of the culture, the process conditions and the characteristics sought in the end product;
(ii) direct inoculation of the bulk starter milk with a concentrated culture purchased from a recognised supplier; and
(iii) direct inoculation of the process milk employing a concentrated culture.

Further details of these alternatives are shown in Table II, and the final choice will depend on an assessment of a number of technical, economic and, perhaps, political factors.

Although the use of direct-to-vat cultures is increasing, particularly for special products, such as 'Cultura', where the ratio between the constituent organisms is critical, many large plants favour the preparation of bulk starters on-site. The reasons for seeking this option have already been mentioned, and despite the obvious attractions, it is an aspect of the overall process that demands especial attention to detail.

The principal requirement is, of course, to achieve sterile conditions, and ensuring that a bulk of several hundred gallons of milk, the headspace above the milk, as well as the associated pipework and fitments are all sterile is a task that calls for the most conscientious attention to detail. In its absence, inhibition of some or all of the inoculated species/strains of starter bacteria may occur as a result of competition from other bacteria, or from the presence of bacteriophage (Cogan & Accolas, 1990).

The introduction of contaminant organisms by the milk—whether fresh or reconstituted, special grade, skim-milk powder—should be nullified by the severe heat treatment that is usually applied to milk for bulk starters, e.g. 95°C for 30 min, but even with this regime, viruses can survive in the head-space of a partially-filled tank. In an attempt to minimise the risk of starter failure, Hup (1985) proposed the following procedure:

— CIP cleaning of bulk tank and pipelines
— fill the tank with skimmed milk
— heat the milk to 95°C
— steam the head-space for some minutes
— maintain a positive pressure in the vat by means of sterile air
— cool the milk to incubation temperature
— inoculate with concentrated starter culture
— incubate to desired acidity
— cool to 4°C for subsequent use, or if the manufacturing programme permits, use immediately without cooling.

This latter approach is certainly employed in some yoghurt factories (Tamime & Robinson, 1985) as a means of ensuring a more rapid fermentation of the process milk.

For some products, such as those that rely on the activities of the mesophilic streptococci, rotation of the cultures or the use of phage-resistant strains can be considered as additional safeguards, as can the employment of 'phage-resistant' or 'phage-inhibitory' media. These

TABLE II

TYPES OF COMMERCIAL STARTER CULTURES AND SOME COMMENTS ON THEIR USAGE

Types of culture	Comments	Storage in creamery	Method of use
Liquid cultures	Expensive in terms of laboratory facilities; requirement for bulk starter and intermediate culture vessels; demanding on creamery personnel; high risk of infection during transfers, e.g. bacteria, yeasts or phage; culture characteristics liable to change with frequent sub-culturing	Refrigeration at 5–7°C	Production of all cultures up to bulk starter
Frozen cultures (a) Short-term	Requirement for central laboratory within company, but also available from independent suppliers; requirement for rapid and reliable transport service; requirement for bulk starter (and perhaps intermediate) culture vessels; low risk of infection if bulk starter facilities are well maintained; less demanding on creamery personnel; improved control of cultures should eliminate undesirable changes	Deep freeze at −20 to −40°C (storage life limited)	Intermediate or bulk starter production

(b) Long-term e.g. Marstar Cultures (Miles Laboratories Ltd) Redi-set Cultures (Chr. Hansen's Laboratorium A/S)	Expensive culture storage facilities; requirement for bulk starter vessels;[a] easy to use; low risk of infection; an extensive spectrum of phage unrelated strains is available, especially of mesophilic streptococci	Liquid nitrogen at $-196°C$ or deep-freeze at $-50°C$ (storage life limited at $-50°C$)	Bulk starter production
Dried cultures			
(a) Freeze-dried (standard) e.g. Driv-Vac Cultures (Chr. Hansen's Laboratorium A/S)	Eliminates need for maintenance of stock cultures; standard culture characteristics; otherwise handled as liquid cultures (see above)	Deep-freeze at -20 to $-40°C$	Renewal of laboratory cultures
(b) Freeze-dried (concentrates) Bulk starter production e.g. Redi-set Cultures (Chr. Hansen's Laboratorium A/S)	Requirement for domestic deep-freeze ($-20°C$) for storage; requirement for bulk starter vessels; low risk of infection and phage unrelated strains available; easy to use; can be employed as direct-to-vat cultures, and have proved especially useful for small-scale production of yoghurt, but procedure would be too expensive for large creameries		Bulk starter production

[a]Some cultures, e.g. Hansen's DVS cultures, can be employed for direct-to-vat inoculation.
Source: Robinson (1981, 1983).

proprietary media are essentially milk-based formulations with growth factors added to encourage the development of the starter bacteria, and phosphate to sequester the free calcium ions in the milk. This latter inclusion reduces the activity of any phages present, because, in general, the extent of phage proliferation is linked to the availability of calcium.

However, although these media have found favour in the cheese industry, most manufacturers of fermented milks rely on an effective heat treatment of the milk supported by a high standard of hygiene.

Nevertheless, even given that the starter milk and the associated plant is sterile, inoculation of the bulk tank calls for extreme caution. A number of systems are available to meet this requirement (Tamime, 1990), and the general aims are:

(i) *To achieve a sterile transfer of the inoculum into the starter tank.* For liquid cultures produced on-site, this transfer may be through a membrane located on top of the tank, and protected from contamination by a solution of hypochlorite. A special hypodermic is used to pierce the membrane, and the transfer of the intermediate culture is mediated either manually from another container or, for larger volumes, by introducing sterile compressed air into the head-space above the intermediate culture. The end result is that the bulk starter milk receives an active culture of the desired species at a rate of around 2% (v/v), and that the bacteria find themselves in an environment that should be totally conducive to their development. The advantage of this approach is that the tank is totally sealed after filling, and hence no air can leave or enter the vessel during the heating/cooling operations. It is, however, a rather inflexible system in terms of alternative forms of inoculum (see Table II), and hence more recent designs allow air to enter or leave the tank through special filters, or maintain a positive air pressure in the head-space through an injection of sterile air. These latter modifications allow a wider range of inoculation ports to be built into the basic design, so that the manufacturer can employ either concentrated frozen or freeze-dried cultures to inoculate the bulk starter milk.

(ii) *To produce a bulk starter with a high number of viable bacteria of the correct species and, in many cases, the correct balance between the species.* In this context, the obvious requirements are:
   (a) the facility to control the conditions of fermentation, e.g. by having a water-jacketed tank for temperature control and/or cooling of the culture at the end of incubation;

(b) the facility to monitor the progress of the fermentation, and more especially, the build-up of lactic acid. Thus, if the mesophilic streptococci are allowed to remain in contact with high levels of lactic acid over a period of time, so their viability will decline, and the same is also true for *Str. thermophilus*. Other starter bacteria, such as *Lac. acidophilus* and *Bif. bifidum*, may be even more sensitive to pH values below 5·0, so that with these species, the on-site production of starters requires extreme care. It is relevant also that these same species are poor competitors in milk, and hence the question of sterility, discussed above, becomes extremely important.

The manufacturers of starter cultures can, of course, circumvent the problem of over-acidity by employing systems that neutralise the lactic acid as it is produced, and a number of approaches have found practical application in this context (Tamime, 1990).

However, in the creamery, sophisticated mechanisms for pH control have found limited application, and an important function of the production manager is simply to halt the fermentation at an early stage. So long as this point coincides with the logarithmic phase of growth for the species in question, then this crude approach is more than adequate. For products requiring a complex culture of several species, then attention may have to be given to the need to produce a number of individual cultures on-site (Schuler-Malyoth *et al.*, 1968), but this demanding procedure is probably best avoided unless commercial constraints dictate otherwise.

Production of the Final Product

The precise lay-out of the system that will be employed for manufacture will depend on many factors, and a number of reviews have assessed the various options available (Tamime & Greig, 1979; Robinson, 1986; Bartholomai, 1987). In addition, the items of plant will have to be selected with due consideration to the physical and other properties of the product for, as shown in Table III, the range of possible retail items is extensive. However, in the present context, the essential requirement is that the retail product should, at the time of consumption, contain an abundant and viable microflora composed of the desired species. In many cases, 'abundant' can be defined as a certain 'therapeutic minimum'/ml of product, and it is vital that manufacturers take the steps necessary to

TABLE III
SOME EXAMPLES OF FERMENTED MILK PRODUCTS AVAILABLE IN THE MARKET PLACE, AND THAT CONTAIN CULTURES WITH ALLEGED HEALTH-PROMOTING PROPERTIES

Type of fermentation	Traditional and/or retail name	Country of origin	Microflora present
I. Lactic acid			
Mesophilic	Taetmjolk Filmjolk Lattfil Langfil	Scandinavia	*Str. lactis* sub-sp. *lactis*, biovar *diacetylactis*, *Leuconostoc mesenteroides* sub-sp. *cremoris*
	Ymer	Denmark	*Str. lactis* sub-sp. *cremoris*, biovar *diacetylactis*
Thermophilic	Yoghurt	most countries	*Lac. bulgaricus*, *Str. thermophilus*
	Bulgarian buttermilk	Bulgaria	*Lac. bulgaricus*
	Liquid yoghurt	Korea	*Lac. bulgaricus* or *Lac. casei* or *Lac. helveticus*
	ACO-yoghurt	Switzerland	Yoghurt culture, *Lac. acidophilus*
	A-38 fermented milk	Denmark	*Lac. acidophilus*, mesophilic lactic acid bacteria

Acidophilus milk	most countries	*Lac. acidophilus*
AB-fermented milk	Denmark	*Lac. acidophilus, Bif. bifidum*
AB-yoghurt	Denmark	As above plus yoghurt culture
Biogarde®		*Str. thermophilus, Lac. acidophilus, Bif. bifidum*
Bioghurt®	West Germany	*Str. thermophilus, Lac. acidophilus*
Bifighurt®		*Bif. bifidum*
Real Active	UK	Yoghurt culture plus *Bif. bifidum*
Mil-Mil E		Yoghurt culture *Bif. bifidum*
Miru-Miru	Japan	*Lac. acidophilus, Lac. casei, Bif. breve*
Yakult		*Lac. casei*
II. Yeast-lactic acid		
Kefir		Refer to text
Kumiss	Russia	Refer to text
Acidophilus-yeast		*Lac. acidophilus*, lactose-fermenting yeasts

Source: Tamime & Robinson (1988).

ensure survival of the constituent organisms at this level. The conditions required to meet this constraint will vary with the species concerned, but as will be discussed in later chapters, the efficacy of the products is dependent, in large measure, on the presence of living bacteria. Indeed, much of the scorn that has, in the past, been heaped upon the concept that fermented milks may have special health-promoting properties has been fuelled by the performance of products containing few, if any, surviving micro-organisms.

Avoidance of these past mistakes is essential for producers and consumers alike, and with the increased knowledge about the species concerned, there is no reason why the present generation of retail items should not meet the minimum requirements in respect of viable starter bacteria. Consumer acceptability, in terms of the organoleptic properties of the product is, of course, of equal concern, because if the vehicle is unpalatable, then clearly there is little point in extolling the virtues of the culture. However, this aspect of promotion should not prove a major obstacle, and there is no reason why many consumers should not now have an opportunity to enjoy some of the products mentioned in Table III and, perhaps, establish for themselves that fermented milks can be a most valuable addition to the diet.

REFERENCES

Alexander, J. G. (1971). *J. Royal College General Pract.*, **21**, 633.
Babel, F. J. (1977). *J. of Dairy Science*, **59**, 200.
Bartholomai, A. (1987). *Food Factories*, Verlag Chemie, Weinheim, Germany.
Chandon, R. C. (1990). ed. *Yoghurt: Nutritional and Health Properties*. National Yoghurt Association, McLean, Virginia, USA.
Cogan, T. M. & Accolas, J.-P. (1990). In *Dairy Microbiology*, 2nd Edn, Vol. 1, ed. R. K. Robinson. Elsevier Applied Science, London, pp. 77–114.
Dellaglio, F. (1988). In *Fermented Milks—Science and Technology*, IDF Bulletin No. 227,7.
FAO/WHO (1967). Expert Committee on Food Additives, FAO, Rome.
Gasson, M. J. (1986). In *Developments in Food Microbiology—2*, ed. R. K. Robinson. Elsevier Applied Science, London, pp. 195–220.
Gilmour, A. & Rowe, M. J. (1990). In *Dairy Microbiology*, 2nd Edn, Vol. 1, ed. R. K. Robinson. Elsevier Applied Science, London, pp. 37–76.
Heap, H. A. & Lawrence, R. C. (1988). In *Developments in Food Microbiology—4*, ed. R. K. Robinson. Elsevier Applied Science, London, pp. 149–86.
Hup, G. (1985). In *Starter Cultures in the Food Industries*. Food Biotechnology Workshop, University College, Cork, p. 7.
Hurst, A. (1983). *Food Science*, **10**, 327.
IDF (1988). *Fermented Milks—Science and Technology* Bulletin No. 227. International Dairy Federation, B-1040 Brussels, Belgium, p. 164.

Kandler, O. (1983). *Antonie van Leeuwenhoek,* **49,** 209.
Kempler, G. M. & McKay, L. L. (1981). *J. of Dairy Science,* **64,** 1561.
Koroleva, N. S. (1988). In *Fermented Milks—Science and Technology,* IDF Bulletin No. 227, pp. 96–100.
Macura, D. & Townsley, P. M. (1984). *J. of Dairy Science,* **67,** 735.
Marshall, V. M. E. (1987). *J. of Dairy Research,* **54,** 559.
McKay, L. L. (1985). In *Bacterial Starter Cultures for Foods,* ed. S.E. Gilliland. CRC Press Inc., Boca Raton, USA, pp. 159–74.
Nahaisi, M. H. (1986). In *Developments in Food Microbiology—2,* ed. R. K. Robinson. Elsevier Applied Science, London, pp. 153–78.
Rašić, J. L. & Kurmann, J. A. (1983). *Bifidobacteria and their Role.* Birkhauser Verlag, Basel.
Reddy, G. V. & Shahani, K. M. (1971). *J. of Dairy Science,* **54,** 748.
Robinson, R. K. (1977). *Nutrition Bulletin,* **4,** 191.
Robinson, R. K. (1981). *Dairy Industries International,* **46**(12), 15.
Robinson, R. K. (1983). In *Biotechnology—3,* eds. H.-J. Rehm & G. Reed. Verlag Chemie, Weinheim, pp. 191–202.
Robinson, R. K. (1986). (ed.) *Modern Dairy Technology,* Vols 1 & 2. Elsevier Applied Science, London.
Robinson, R. K. (1988). *Dairy Industries International,* **53**(7), 15.
Robinson, R. K. & Tamime, A. Y. (1990). In *Dairy Microbiology,* 2nd Edn, Vol. 2, ed. R. K. Robinson. Elsevier Applied Science, London, pp. 291–344.
Schuler-Malyoth, R., Ruppert, A. & Muller, F. (1968). *Milchwissenschaft,* **23,** 554.
Tamime, A. Y. (1990). In *Dairy Microbiology,* 2nd Edn, Vol. 2, ed. R. K. Robinson. Elsevier Applied Science, London, pp. 131–201.
Tamime, A. Y. & Deeth, H. C. (1980). *J. of Food Protection,* **43,** 939.
Tamime, A. Y. & Greig, R. I. W. (1979). *Dairy Industries International,* **44**(9), 8.
Tamime, A. Y. & Robinson, R. K. (1985). *Yoghurt—Science and Technology.* Pergamon Press, Oxford.
Tamime, A. Y. & Robinson, R. K. (1988). *J. of Dairy Research,* **55,** 281.

Chapter 3

THE THERAPEUTIC EFFECTS OF VARIOUS CULTURES—AN OVERVIEW

LIVIA ALM

Medical Nutrition Department, Karolinska Institute, Stockholm, Sweden

Fermentation is one of the oldest and most widespread methods of preserving food, particularly milk. Practically all nations have some traditional type of fermented milk product made by the action of lactic acid-producing micro-organisms. Fermentation has often been something of a tradition, and the art of preservation was handed down from one generation to the next. From prehistoric times man learned to use milk as food and, although the origin of the art of preserving dairy products by lactic acid fermentation is lost in antiquity, the biochemical and microbiological knowledge of fermentation is of comparatively recent date.

New, in this context, are the therapeutic fermented milk products, where the micro-organisms used are usually of special origin, with carefully investigated properties. It is required of therapeutic lactobacilli that they have not only biochemical and biological effects on milk nutrients, but also physiological and therapeutic effects on the consumer.

METCHNIKOFF AND HIS AUTO-INTOXICATION THEORY

There are scientists who claim a longer life expectancy is to be found in those societies where fermented milk products are basic elements of the diet. Others see nothing more than good staple foods, with high nutritional qualities equal to those of non-fermented milks.

Ilja Metchnikoff (1908) was the first to advocate the consumption of fermented milk, i.e. yoghurt, as beneficial to health. The scientific basis for his recommendation has been a matter of much controversy. Metchnikoff's auto-intoxication theory claims that 'the human body is slowly being poisoned and its resistance weakened by the action of a wrong type of intestinal flora. Death will come more rapidly to a heavy meat eater; meat putrifies, whereas milk does not, because of its content of lactose, from which the lactobacilli, if present, produce lactic acid, which together with other metabolites, can control the number of disease-causing microorganisms in the milk and in the human organism'.

Fermented milk alone, however, does not prolong life—there is so much more that has to be considered. Metchnikoff himself consumed large quantities of yoghurt, but he started the supplementation of his diet with lactobacilli too late in life, and died only 71 years old.

In spite of much controversy and discussion among scientific groups since the beginning of this century, it is agreed upon that our well-being is dependent on a well-functioning intestinal flora, and that the presence of lactobacilli in the gastro-intestinal tract is essential for a healthy life. If disturbed, altered in composition or minimalized, the intestinal flora must be quickly restored and normalized. This can only be accomplished by administration of the correct types and amounts of lactobacilli with some type of carrier, usually cows' milk, or some other substrate. Lactobacilli can be administered in many forms, but it is important to provide viable cells that are able to survive passage through the gastro-intestinal tract.

Fermented milk products can be divided into traditional, home-made products, where the fermentation is carried out by the lactobacilli present in the environment (these being mesophilic or thermophilic) without knowing the exact composition of the starters; and industrially-made products, where the micro-organisms are carefully chosen from a culture bank, having been investigated and propagated especially for the purpose of producing a typical product of high quality and good sensoric properties. These latter milks resemble the home-made types, but enjoy a more controlled outcome, higher hygienic quality and longer keeping time.

TYPES OF MILK USED

All mammalian milk can be used for the production of fermented milk products, but most usual is cows' milk as such or supplemented with growth factors. Milk differs in composition, not only among various species but also within species and individuals. Some of its constituents, such as milk fat, with its specific fatty acid composition, lactose and casein are not found elsewhere in nature. The nutritional value of cows' milk is generally accepted as being the highest, and provides most nutrients in a bio-available form.

MICRO-ORGANISMS USED IN FERMENTATION

Starters used in fermentation can be divided into two main categories. Technical starters, which are good fermenters, use milk nutrients for their own metabolism and produce palatable products, but do not resist the acid environment of the stomach or bile; and therapeutic starters, additionally highly acid and bile tolerant, which can survive passage through the intestinal tract and influence the health, well-being and longevity of the consumer.

Lactic acid bacteria are Gram-positive, usually non-motile, non-sporulating rods and cocci, catalase-negative, obligate fermenters. The utilization of hexoses, such as glucose or fructose, allows additional differentiation into homo- and heterofermentative species. The homofermentative species produce mainly lactic acid, the heterofermentative species also produce volatile fatty acids, ethanol and carbon dioxide. Classification is usually based upon morphology (rods or cocci), growth temperature requirements (thermophilic or mesophilic starters), energy utilization (homo- or heterofermenters) and the production of aroma components.

TABLE I
PRODUCT TYPES, LACTIC STARTERS USED AND THEIR PROPERTIES

Products	Starters used	Properties
Low fat buttermilk	Streptococci and *Leuconostoc* spp.	Technical
Buttermilk	Streptococci and *Leuconostoc* spp.	Technical
Fermented cream	Streptococci and *Leuconostoc* spp.	Technical
Ropy milk	Streptococci	Technical
Kefir	Lactobacilli, Streptococci and yeasts	Technical
Kumyss	Lactobacilli, Streptococci and yeasts	Therapeutic
Acidophilus milk	Lactobacilli	Therapeutic
Bifidus milk	Bifidobacteria	Therapeutic
Yoghurt	Lactobacilli and Streptococci	Therapeutic

Many new types of products with documented properties are made by the use of combinations of the above-mentioned starters.

FERMENTATION

When the fermentation process is carried out in a modern dairy, it can be characterized as a controlled chemical and biological food preservation process. As a result of fermentation, conditions are provided for an incomplete metabolism of the components of milk (a predigestion that is considered as beneficial) and the production of intermediates, the most important of these being lactic acid and other organic acids, as well as many other organic compounds with the ability to control the behaviour of various micro-organisms and, to some extent, also affect the consumer. The lowered pH in all fermented milk products compared with milk not only retards the growth of undesirable micro-organisms in the product, but also provides a pleasant taste and better keeping properties.

Lactobacilli use the nutrients in milk for their own metabolism and growth, and multiply from one to ten million cells per millilitre to one thousand million cells per millilitre. These micro-organisms are present in the fermented milk, not only as viable cells, but also as autolysing cells which give rise to primary and secondary metabolites and enzymes that they may have produced during fermentation and continued to produce during storage.

The nutritional value of a food depends not only on the nutrient content but also on the bio-availability of the nutrients. Fermented milk products generally contain the same amounts of nutrients as the milk from which they are made, but the fermentation has made some of the nutrients more available for absorption. Interaction between milk nutrients and starters results in the following changes:

Effect on Proteins

Proteins in milk show only relatively slight biochemical and nutritional changes after fermentation, except for potential curd size, which is reduced. Curd particle size is recognized as an important factor in the *in vivo* digestibility of proteins, and milk types which yield small curd particles are believed to be well tolerated and utilized by infants, children and adults. The beneficial effect of predigested milk products in the diets of patients, especially with protein-energy malnutrition and gastrointestinal disturbances, has been noted, and X-ray studies of gastric emptying rates in piglets showed that fermented milk products leave the stomach faster than unfermented milk (Alm, 1983a).

Milk Fat Degradation

Lactobacilli exert low lipase activity, but milk fat is partially degraded and its digestibility enhanced (Blanc, 1973; Rašić & Kurmann 1978). The presence of most of the fat-soluble vitamins represents a valuable contribution to the physiological value of milk fat in fermented milk products.

Effect of Lactose Content

Lactose generally decreases in fermented milk products. The decrease is 30–50%, which is very important with regard to the problems of lactose intolerance/maldigestion. The intestinal flora metabolizes lactose, forming lactic acid and other volatile and non-volatile compounds. In small amounts, lactose is beneficial to the human, but in larger amounts it will cause irritation of the intestinal mucosa leading to abdominal cramps and diarrhoea. In lactose-intolerant children, lactose may even be toxic and damage the intestinal mucosa giving rise to symptoms similar to those of coeliac disease. Steatorrhoea has also been reported in individuals with low lactase activity. The enzyme, lactase, was present in *Lactobacillus bulgaricus* and *Streptococcus thermophilus* isolated from fresh yoghurt. The activity of the lactase increased during incubation, reaching its highest activity after 4 h at 43°C. Lactase is bound in the cells, and was shown to be released by simulated gastric digestion (Breslaw & Kleyn, 1973). Therapeutic lactobacilli, in general, contribute active lactase to the fermented milk product.

Lactic Acid Production

Lactic acid is the characteristic substance in all fermented dairy products. The physiological role of the lactic acid isomers $L(+)$, and $D(-)$ and DL in man and animals has been investigated (Cohen & Woods, 1976). Both isomers are absorbed from the human intestinal tract, although the rate of metabolism of the $D(-)$ isomers is considerably slower. It is important to remember that the World Health Organisation recommends that products containing $D(-)$ and DL lactic acid should be restricted in infant feeding. Lactic acid and the remaining amount of lactose in fermented milks affect calcium absorption favourably (Dupuis, 1964). Higher calcium retention has been found in rats consuming fermented milk than in those consuming non-fermented milk.

Lactic acid in fermented milk products is desirable for many reasons. As a natural preservative it makes fermented milk products biologically safe, even in hot and contaminated environments. Furthermore, lactic acid makes components of milk more easily digestible. Even achlorhydric individuals (a common disturbance in the aged) can tolerate normal amounts of fermented milk because of its lactic acid content and low pH.

Volatile Fatty Acid Production

Volatile fatty acids in fermented milk products influence the sensoric properties of the products, and may also be important from the nutritional point of view. Acetic acid in relatively large quantities is produced both by *Lactobacillus acidophilus* and bifidobacteria. The antimicrobial properties of fermented milks products are derived primarily from the lactic acid content, but are intensified by the other organic components (Speck, 1978). It is not known whether acid production continues to a significant degree in the intestine. If it should, it will be produced by *Lac. acidophilus* and, at a higher level, by bifidobacteria (Alm, 1982).

Vitamins

Fermentation affects most B-vitamin levels in the final product. While many lactobacilli require B-vitamins for growth, several lactobacilli are capable of synthesizing B-vitamins in excess. During fermentation there is a dramatic increase in cell population, which continues, even if very much slower, during storage.

A very dramatic increase in folic acid and a decrease in vitamins B_{12} has been reported (Reddy *et al.*, 1976). An important question concerning B-vitamins in fermented milk products is their bio-availability. Thus B-vitamins function as integrated parts of protein and enzyme systems and, therefore, may occur in both free and protein-bound forms in the cells. It is not known whether or where 'cell-bound vitamins' are made available to the host. It may be speculated that one way in which lactobacilli may influence the metabolic activities of the normal intestinal flora is by supplying it with B-vitamins.

Production of Biocines

There is now conclusive evidence that lactobacilli produce substances that inhibit the growth of pathogens *in vitro* (Shahani & Ayebo 1976; Stadhouders *et al.*, 1978). The use of lactobacilli as dietary supplements has been found to alleviate intestinal infections in humans and also in animals (Shahani & Ayebo, 1980). However, it is not clear whether the lactobacilli control the infection, or whether they accelerate the establishment of a normal flora after the body's defence mechanisms, such as the immune system, have controlled the infection. Among the various by-products formed by the lactobacilli during their growth, hydrogen peroxide, in spite of being produced in extremely small amounts and being transient and difficult to measure, has been found to inactivate harmful bacteria *in vitro* (Gilliland & Speck, 1975).

Other biologically active compounds are biocines, synthesized as secondary metabolites by various lactic acid bacteria—*Lactobacillus plantarum* produces lactolin (Kodama, 1952); *Lactobacillus brevis* produces lactobrevin (Kavasnikov & Sodenko, 1967); *Lactobacillus bulgaricus* produces bulgarican (Reddy & Shahani, 1971); and *Lactobacillus acidophilus* has been reported to produce several inhibitors, such as acidophilin (Vakil *et al.*, 1965), lactocidin and acidolin (Hamdan *et al.*, 1973). Biocines exert bacteriocidal and bacteriostatic effects against disease-causing micro-organisms. The action of lactobacilli and the biocines produced by them has also been described as antitumorigenic and anticholesterolemic (Shahani & Chandani, 1979). Consequently, some types of fermented milk product containing special types of lactobacilli may exert a therapeutic effect in the intestinal tract, especially following colonization or, at least, viable passage in large numbers.

Thus, the nutritional and physiological effects of milk fermentation are the following:

— the digestibility and absorption of milk proteins is enhanced
— the digestibility of milk fat is improved
— the lactose content is reduced
— better tolerance of milk by lactose maldigestors, or individuals suffering from low HCl secretion and high pH in the stomach
— retention of certain minerals, such as calcium, is enhanced
— the content of certain B-vitamins is enhanced
— fermented milk products provide the consumer with high numbers of viable and non-viable cells of starter origin, and with all the metabolites that they produce during fermentation and storage.

HUMAN FLORA AND ITS ROLE

Colonization in Early Life

During prenatal life, the foetus lives in a sterile environment, but the infant at birth is contaminated with both the mother's flora and bacteria from the environment, which results in a rapid colonization of the intestinal tract.

Bifidobacteria appear in the stools of breast-fed babies from 2 to 5 days after birth, with the establishment of a relatively stable microflora in the colon within several days. By the end of the first week, biofidobacteria become the predominant micro-organisms of the faecal flora and rise to about 99%. The pH values of the stools range from 5·0 to 5·5, or only slightly higher, in the breast-fed infants.

Weaning causes a gradual change in the composition of the faecal flora.

The faecal flora of bottle-fed infants resembles that of older children and adult individuals and, although bifidobacteria and lactobacilli remain the principal part, they decrease in numbers by 1 or 2 logs.

The pH value increases to about 7·0 or even higher, which indicates the presence in high numbers of putrefactive micro-organisms.

Establishment of Adult Flora

Normally, both non-pathogens and potential pathogens (actually unwanted micro-organisms) are present in the stomach, ileum, caecum and colon, where they show much variety concerning their requirements for life. The regulating mechanisms concerning the numbers and species composition of the flora are very complex. More than 40% of the faeces mass are intestinal micro-organisms comprising between 400 and 500 different species. The intestinal flora is a dynamic organ, with a very rapid turnover, containing ten times as many cells as the human body (Luckey, 1979). Micro-organisms in the gastro-intestinal tract of normal adults are always colonizing the same region, reach high numbers and remain at that number as long as the individual is healthy, and has a normal nutritional status. The physiology, age and diet of the host, among other parameters, influence the distribution of various species and biotypes of lactobacilli and bifidobacteria in the large intestine. The population of lactobacilli in the gastro-intestinal tract of adults is stable, although changes may be caused by various internal and external factors.

The stomach of normal, fasting individuals is nearly sterile—less than one thousand organisms per millilitre—and the pH is usually less than 3·0, while in the achlorhydric stomach associated with pernicious anaemia or the effect of ageing, the pH is often greater than 6·0. The bacteriostatic effects of gastric juice is partly due to the low pH; no such effect has been observed above a pH of 4·0. The bacteriostatic effects of gastric juice may be influenced by the following factors: the number of micro-organisms ingested at one time; the rate of gastric emptying; the physical protection of micro-organisms by ingested food; and buffering of the stomach contents.

The upper small intestine may contain up to 10^4 micro-organisms per millilitre of content. These include streptococci, lactobacilli, yeasts and small numbers of bifidobacteria and bacteroides. The lower parts of the small intestine harbour a rich microbial population of up to 10^5–10^7 organisms per millilitre of content, consisting of streptococci, lactobacilli, bifidobacteria, bacteroides, enterobacteria and yeasts. The distribution of the bacterial flora in the small intestine is influenced by the effect of gastric acidity. Disorders affecting intestinal motility usually lead to 'bacterial overgrowth' of the small intestine. Other factors which may influence the composition and quantity of the microbial population are the type of diet and the rate of gastric emptying. Intestinal immunoglobulins, lysozyme and bile salts may selectively affect the intestinal bacteria.

The large intestine extends from the ileum to the anus, and consists of the caecum, colon and rectum. Food residues take 18–68 h to pass through. The microbial population of the large intestine is qualitatively similar to that of faeces, upon which most investigations have been conducted for practical reasons. Up to 40% of the faecal mass consists of bacteria, and the number of cultivable micro-organisms ranges from 10^{10} to 10^{11} g^{-1} wet weight.

Anaerobic micro-organisms are the major components and comprise more than 90% of the micro-organisms. Lactobacilli, enterococci and coliforms form less than 1–5% of the cultivable flora.

In adults, influences, such as disturbances of gastric function, disorders of intestinal motility, stagnation of the intestinal content due to blind loops, impaired intestinal motility, e.g. diverticulosis, regional enteritis (Crohn's disease), scleroderma, X-ray irradiation of the abdomen, intestinal strictures, gastrocolic and enterocolic fistulae, liver cirrhosis, and disorders of the immunological system, or simply extreme situations of stress, cause changes in the composition of the gut flora.

Protective Functions of the Normal Flora

The gut micro-organisms may inhibit the growth of invading pathogens by the production of organic acids, particularly volatile fatty acids, by the production of bacteriocines, and by the deconjugation of bile salts. Bacteriocines are produced by a number of members of the intestinal flora, including the endogenous strains of lactobacilli. For example, *Escherichia coli* is able to produce colcines and microcines, and five different enterocines may be produced by enterococci. If disturbed or minimized, the gastro-intestinal flora must be replaced as soon as possible. The easiest way to accomplish this is by the administration of viable cells of lactobacilli.

THERAPEUTIC LACTOBACILLI

There are hundreds of strains of lactobacilli and bifidobacteria which can be used in the fermentation of milk. However, if the organisms should be able to exert prophylactic or therapeutic effects on the consumer, it is required that strains possess some very important properties. With the microbiological technology available today, cells of selected cultures can be prepared for diet supplementation that possess high levels of viability and special metabolic activities. Qualities such as host specificity, compatibility with the host (attachment), beta-galactosidase activity, and ecological interactions with other species within the intestinal microflora, may be important and have to be considered.

Survival through Gastric Acidity

Therapeutic lactobacilli, streptococci and bifidobacteria must have the ability, to survive to a certain extent, the acidity of the gastric juice, and pass in a viable state to the small intestine region. Administration of large numbers of lactobacilli and bifidobacteria may, consequently, increase the number of survivors, but various strains may differ in acid tolerance and survival. However as the transit time of 'predigested' fermented milk foods through the stomach is shorter than for ordinary food, and it has been shown that the administration of serologically and biochemically defined strains of lactobacilli and bifidobacteria leads to the appearance in high numbers of the same strains in the faeces (Alm et al., 1988).

TABLE II
SOME OF THE MAIN FACTORS INFLUENCING THE COMPOSITION OF
THE INTESTINAL FLORA

(A) Host factors	Hydrochloric acid secretion
	Enzymes and bile salts
	Immune mechanisms
	Peristalsis
	Physiological status of the host
	Age
(B) Environmental factors	Infection
	Diet
	Drugs
(C) Bacterial interaction	Antagonism
	Symbiosis

Tolerance of Bile Acids

The levels of faecal bile acids are influenced by the fat content of the diet, and fatty acids may increase the inhibitory effects of bile acids towards lactobacilli and bifidobacteria. Therapeutic lactobacilli should be able to survive relatively high concentrations. Most strains of lactobacilli and bifidobacteria deconjugate bile salts to free bile acids, which are more inhibitory to susceptible bacteria than the conjugated forms (bile salts).

Ability to Produce Volatile Acids *in situ*
The presence of micro-organisms that can produce volatile fatty acids *in situ* is highly desirable. This can be accomplished by the promotion of growth of and/or ingestion of lactobacilli and bifidobacteria. Organic acids produced by lactobacilli and bifidobacteria play a major role in the antagonistic effects recorded both *in vitro* and *in vivo* against many unwanted micro-organisms (Mayer, 1966). Organic acids present in the free (undissociated) forms may have a direct toxic effect on pathogens, but they differ in their anti-microbial effects. Acetic acid, for example, has a stronger ability to inhibit microbial growth than lactic acid. Unlike the homofermentative lactic acid bacteria, bifidobacteria and some lactobacilli (*Lac. acidophilus*) produce more acetic acid than lactic acid from fermentable carbohydrate, and also produce a small amount of formic acid. The acid-producing ability, and the proportions of the fermentation products, may vary among different strains within the same species. By their presence, some gut micro-organisms, e.g. lactobacilli and bifidobacteria, also stimulate intestinal peristalsis, thus alleviating the prevalence of constipation and achieving, indirectly, the inactivation of invading pathogens.

Normally, the indigenous gut flora acts synergistically with its host's immunological system in protecting against intestinal pathogens.

Required properties of therapeutic lactobacilli are

— relatively fast growth in the medium—usually milk
— presence in high cell numbers at the end of fermentation
— no production of any harmful components
— the ability to pass through the GI tract in a viable state
— the ability to multiply in the GI tract, produce organic acids and other biologically active compounds *in situ* and, thereby, control the microbial environment.

Unfortunately, the lactobacilli in most conventional fermented milk products do not possess the mentioned properties. These will not have the expected preventive or therapeutic effects. Exceptions are the carefully investigated viable cells of lactobacilli and bifidobacteria.

EFFECTS OF THE ADMINISTRATION OF LACTOBACILLI

Biochemical Effects
Lactobacilli improve the absorbability of nutrients, improve lactose intolerance symptoms, metabolize some types of drugs, deconjugate bile acids, reduce the level of serum cholesterol and reduce the risk of colon cancer, etc.

Physiological Effects
Lactobacilli improve intestinal motility, stimulate the immune system, reduce the incidence of tumours.

Antimicrobial Effects
Lactobacilli act in a protective way, create an antagonistic environment against pathogens and stabilize the composition of the intestinal microflora.

Competitive Colonization
Lactobacilli act by blocking the receptors or the adhesion sites of pathogens; act by inactivating the effects of enterotoxins, by forming a mixture of non-pathogens and non-antibiotic resistant floras, especially with *E. coli*.

PREVENTION AND THERAPY BY THE USE OF LACTOBACILLI

The concepts of Metchnikoff (1908) are finally being confirmed both experimentally and clinically. The influence of the balance of the normal flora of the intestinal tract, on the health and well-being of the host, is now well documented. Recent research with continuous feeding of lactobacilli shows positive results, especially with aged individuals, so long as adequate numbers of viable organisms are fed to hosts (Shahani & Chandani, 1979; Blanc, 1981; Alm *et al.*, 1983).

Re-establishment of the Intestinal Flora by the use of Lactobacilli and Bifidobacteria
The normal gastro-intestinal flora can be re-established, when influences causing abnormal distribution of intestinal micro-organisms are eliminated, and this re-establishment can be accelerated by the administration of lactobacilli, such as *Lac. acidophilus* or/and bifidobacteria.

In the following section, some of the more important applications of culture administration for dietary, prophylactic and therapeutic uses will be reviewed, namely:

— infantile diarrhoea
— traveller's diarrhoea
— antibiotic-induced diarrhoea
— stress situations
— cancer therapy
— colon cancer
— toxic amines
— carcinogens
— constipation
— cholesterol reduction
— lactose intolerance

Infantile Diarrhoea
Investigations into the role of lactobacilli in infantile diarrhoea (Bianchi-Salvadori 1981) provide strong evidence that the lactobacilli in yoghurt are able to pass through the gut in a viable state, and act to increase the number of autochthonous lactobacilli (*Lac. acidophilus*) and bifidobacteria. These lactobacilli and bifidobacteria may prevent the 'overgrowth' of harmful micro-organisms in the intestine following antibiotic therapy, or other treatments. It has been shown that the oral administration of milk preparations containing *Lac. acidophilus* and *Bif. bifidum* to an infant, following the discontinuation of penicillin treatment, increased the population of lactobacilli and bifidobacteria in faeces, and suppressed the growth of *Candida albicans*.

Traveller's Diarrhoea
Travellers all over the world are plagued by many kinds of gastro-intestinal disturbance, and some of these can turn out to be long lasting infections, resulting in a 'carrier state'. After the first acute phase of illness, a high number of individuals, although not ill, continue to excrete pathogens in the faeces. This situation may cause social, psychological and economical problems for both the individual and for the society. An investigation (Alm, 1983b) has shown that administration of *Lac. acidophilus* in high cell numbers per day could result in a significantly shortened carrier time. The administration of lactobacilli prior to, during and after a stay in heavily contaminated areas, can protect an individual from disease caused by intestinal pathogens.

Preventing Side Effects in Antibiotic Treatment

Therapy with any antibiotic, particularly long-term and involving oral administration, is liable to alter the balance of the intestinal flora. Powerful antibiotics, such as clindamycin, lincomycin and others, may cause drastic changes. Therapy with clindamycin may eliminate most of the anaerobic micro-organisms, such as bifidobacteria, lactobacilli, bacteroids, anaerobic cocci with the exception of eubacteria and clostridia, while lincomycin completely eliminates the normal faecal flora more or less persistently. Antibiotic therapy has often caused intestinal distress subsequent to controlling the actual causes of infections. Coliforms, in particular, are able to proliferate and produce the cells and gas that cause diarrhoea and flatulence. The administration of cultivated intestinal lactobacilli has been found to re-establish the intestinal flora, and bring relief from intestinal distress.

Changes in the normal intestinal flora resulting from antibiotic therapy emphasize the importance of re-establishing, as soon as possible, the normal microbial balance. The dietary administration of lactobacilli, preferably in conjugation with bifidobacteria, may help in the regeneration of the normal gut flora composition and function.

Cancer Therapy

Neumeister (1969) investigated variations in the normal bacteria flora of the small intestine during and after irradiation therapy in 500 patients with cancer. In order to eliminate some of the disturbances, 162 patients were orally administered a mixture of micro-organisms consisting of non-pathogenic *E. coli, Lac. acidophilus* and *Bif. bifidum*, together with pancreatic enzymes. The administration of the mixture started at the same time as the gamma-ray treatment, and continued during the whole treatment time. Using gamma-ray therapy and supplementation with lactobacilli, the frequency of diarrhoea and other side effects of the treatment was reduced from 61% to 21%, which are highly significant values. The administration of therapeutically-active lactobacilli during cancer treatment is highly recommended.

Effect on Colon Cancer

Epidemiological studies have shown that the incidence of colon cancer is higher in populations consuming 'western' diets than in those customarily on vegetarian diets. In one such study involving populations in Copenhagen (high risk) and Kuopio, Finland (low risk), there was a four-fold variation in the incidence of colon cancer.

Higher intakes of dietary fibre and milk in the 'low incidence' area indicated a protective effect that was unrelated to transit times. An interesting observation was that, in the 'low incidence' area, faecal samples showed a significantly higher population of lactobacilli. The importance of intestinal lactobacilli in the aetiology of colon cancer has been studied by Goldin *et al.* (1976, 1977, 1978, 1980). They showed that feeding rats mostly beef caused much higher levels of bacterial nitroreductase, azoreductase, and beta-glucuronidase, than when the diet was high in vegetable and grain. It was postulated that the high beef diet favoured the establishment of an intestinal microflora with the metabolic potential to convert procarcinogens to carcinogens. However, when the animals obtained the same diet but supplemented with viable lactobacilli, the bacterial enzymes associated with lactobacilli (*Lac. acidophilus*) resulted in significant decreases in the harmful, enzyme activities of the faecal bacteria; thirty days after discontinuing the supplements, the potentially damaging enzyme activities returned to previous levels.

Effects on the Production of Toxic Amines
The beneficial roles of lactobacilli and bifidobacteria in preventing the formation of amines have been pointed out. The higher acidity of the low intestinal contents may indirectly prevent the production of toxic amines from the amino acids by putrefactive bacteria, e.g. clostridia, coliforms, etc.

Effect on Carcinogens
The presence of nitrates and nitrites in foods, and the conversion of nitrites into nitrosamines, has been the subject of much controversy. These compounds occur naturally (White, 1975) in many foods, or are formed in the intestine by certain micro-organisms. There is evidence that lactobacilli in the intestinal microflora biodegrade nitrosamines (Hill *et al.*, 1970; Rowland & Grasso 1975); lower concentrations of amines were found in animals being fed lactobacilli (*Lac. acidophilus*). In piglets receiving lactobacilli, diarrhoea was of shorter duration and less severe than in the control animals. Evidence exists showing that lactobacilli (especially *Lac. casei*) and related micro-organisms biodegrade nitrates and nitrites (Goodhead *et al.*, 1976; Przybylowski *et al.*, 1978).

Effects on Constipation
The administration of the lactobacilli to severely constipated geriatric patients has a beneficial effect on their bowel movement and stool

frequency; this has been shown in several studies (Tanaka et al., 1982; Alm et al., 1983). Regular bowel movements and a properly functioning intestinal tract are important for every individual, but are of the utmost importance in old age. The lactobacilli/lactic acid-containing products with their anti-bacterial properties, help to correct both constipation and diarrhoea as well as intestinal intoxications due to colonic dysfunctions. The consumption of predigested fermented milk products has many advantages for aged individuals. In addition, osteoporosis, frequently prevalent in old age, may be positively influenced by the ingestion of fermented milks with their easily absorbable calcium and phosphorus contents.

Effects on Lipids and Cholesterol
Results from several studies indicate that the reduction of total plasma cholesterol can lower the incidence of coronary heart disease and several findings indicate that the consumption of lactobacilli reduces the serum cholesterol level (Mann & Spoerry, 1974; Mann, 1977). Sterols and lipids are altered markedly by intestinal micro-organisms, and bile acids are deconjugated and dehydroxylated. Possibly the deconjugation of bile acids by intestinal lactobacilli may function in the control of the microbial flora (Shimada et al., 1969; Binder, 1973; Gilliland & Speck, 1977). The influence of the intestinal flora on sterol and bile acid metabolism appears to be advantageous to the host and may be of significance in human nutrition in preventing the accumulation of excessive cholesterol (Coates, 1975). The feeding of a milk formula supplemented with lactobacilli to infants was shown to result in lower levels of blood cholesterol than when milk without lactobacilli was fed to them. Similar observations were reported on piglets that were fed cholesterol-containing diets (Mott et al., 1973).

Lactose Intolerance
Cultures used in conventional fermented milk products (buttermilk, ropy milk, kefir) are unable to survive passage through the gut in noteworthy numbers, and thus benefits for lactose maldigestors from such products are transitory, i.e. during the time lactase from the cells remains in the small intestine (Alm et al., 1988). Several studies have been conducted to determine the value of lactobacilli (*Lac. acidophilus*) in reducing lactose maldigestion.

In one study, six adult, caucasian females (with a history of lactose intolerance) were given 3·5 pints per day of sweet acidophilus milk; six other females were given the same amount of milk, but without the culture. After one week, the groups switched and consumed the other milk type for the following week. The usual lactose challenge (50 g lactose in water) was administered to each person and blood glucose measurements performed. The results (Speck, 1978) indicated that the increased numbers of *Lac. acidophilus* in the intestine had enabled lactose maldigestors to handle the lactose satisfactorily. The data also showed that, without replenishing the supply of *Lac. acidophilus* for one week, the lactose-intolerant individuals were no longer able to digest lactose adequately. The supplementation with lactobacilli must be continued in order to control the maldigestion.

It has been shown (Goodenough & Kleyn, 1976) that rats fed natural yoghurt containing viable cultures were able to digest lactose more efficiently than animals fed other experimental diets, including pasteurized yoghurt. It was concluded (Kilara & Shihani, 1976) that lysis of the culture cells releases lactase, which hydrolyses the lactose contained in the consumed fermented product. Evidence points to the activity of the lactase from cells of the culture enabling lactose-intolerant persons to consume milk products with no subsequent ill-effects.

Studies were conducted (Gilliland, 1981) on lactose maldigestors using breath hydrogen level as a measure of lactose maldigestion; subjects consumed 5 ml milk/kg body weight twice daily. Breath hydrogen analyses were conducted on the subjects before and after a seven-day period, during which they consumed the milk without lactobacilli. Those drinking milk plus lactobacilli gave significantly lower breath hydrogen values on Day 7, further confirming that many persons unable to digest lactose properly can be assisted by the use of *Lac. acidophilus* as a dietary supplement.

Until recently, lactase deficiency among the Swedish population was thought to be fairly low. However, the number of adult immigrants has increased during the last decade, and so has the number of adopted children from countries where the majority of the population is lactose-intolerant. For these and other reasons, it is of interest to reduce the concentration of lactose in milk. This reduction is easily accomplished by fermentation, and lactose-intolerant individuals can usually consume fermented milk products without the customary symptoms developing (Gallagher *et al.*, 1974; Alm, 1982); the starter cultures, which contain lactase, hydrolyse lactose into its two simple sugar components, glucose and galactose.

Disturbances in Stress Situations
The effect of stress on the composition of the intestinal flora has been investigated (Lizko et al., 1984). It was found that lactobacilli decreased when an individual is under stress. Some individuals experience diarrhoea, others only have minor symptoms, but if stress periods are of long duration many types of gastro-intestinal disorder can occur.

The quantitative composition of the gut microflora, with regard to lactobacilli number and type of, before and after cosmic flights of different duration were investigated (Smirnov et al., 1982; Lencner et al., 1984). Due to emotional stress before take-off, the flora of the cosmonauts showed changes regarding the lactobacilli. After a flight, the number of potentially pathogenic enterobacteria was markedly increased, and the lactobacilli almost eliminated. It was also shown that the administration of products containing lactobacilli could be used with success in order to minimize the disturbances caused in cosmonauts. Everyday stress situations can also cause alterations in the composition of the gastro-intestinal flora, and these too can be overcome by a daily supplementation with foods containing lactobacilli.

IN CONCLUSION

The consumption of fermented milk products with therapeutic lactobacilli is suggested for situations of gastro-intestinal and other types of infections; reduced secretion of gastric and intestinal juices; lactose intolerance symptoms; most stress situations; side effects of long-term use of drugs; liver and bile malfunctions, but also in slimming programmes; repair of skin injury; eczema; oral infections and many other diseases.

Lactobacilli appear to be among the friendliest micro-organisms that maintain interactions with man and animals. Great benefits to the host can be derived if micro-organisms are provided of the right type and in the right numbers, together with other food ingredients, to promote a well functioning gastronomical tract. It is important to remember that lactobacilli:

— are essential in the newborn;
— improve the nutritional status of the adult individual; and
— reduce the number of potentially pathogenic micro-organisms and act protectively toward the host, and in doing so, lactobacilli prolong a healthy life.

REFERENCES

Alm, L. (1982). The effect of fermentation on nutrients in milk and some properties of fermented liquid milk products. Thesis, Department of Medical Nutrition, Karolinska Institute, Stockholm.
Alm, L. (1983a). Kieler Milchwirtschaftliche Forschungsberichte. 3, 329–32.
Alm, L. (1983b). In *Proceedings of 7th International Symposium on Intestinal Microecology*, Boston, 1982. *Intestinal Microecology*, 7(3/4) 13–14.
Alm, L., Humble, D., Ryd-Kjellen, E. & Betterberg, G. (1983). Suppl. Näringsforskning, XV Symposium Swedish Nutrition Foundation, pp. 131–8.
Alm, L., Leijenmark, C.-E., Persson, A.-K. & Midtvedt, T. (1988). *Wenner-Gren International Symposium Series, Vol. 52. The Regulatory and Protective Role of the Normal Microflora*, pp. 293–7.
Bianchi-Salvadori, B. (1981). Symp. Int. sur les effets nutritionnels de la flore digestive, Paris, pp. 67–87.
Binder, H. J. (1973). *Gastroenterology*, 65, 847–50.
Blanc, B. (1973). *Schweiz. Milch-Ztg.*, 99, 463–5, 472, 476.
Blanc, B. (1981). Symp. Int. sur les effets nutritionnels de la flore digestive, Paris, p. 118–49.
Breslaw, E. S. & Kleyn, D. H. (1973). *J. Food Sci.*, 38, 1016.
Coates, M. E. (1975). *Bibl. Nutr. Dieta.*, 22, 101.
Cohen, R. D. & Woods, H. F. (1976). *Clinical and Biochemical Aspects of Lactic Acidosis*. Blackwell Scientific, Oxford.
Dupuis, Y. (1964). *Ann. Bull. Int. Dairy Fed.*, Part III: 36–43.
Gallagher, C. R., Molleson, A. L. & Coldwell, J. R. (1974). *J. Am. Diet. Assoc.*, 65, 418–19.
Gilliland, S. E. (1981). *J. Food Protection*, 42, 164–7.
Gilliland, S. E. & Speck, M. L. (1975). *J. Food Protection*, 38, 903.
Gilliland, S. E. & Speck, M. L. (1977). *J. Food Protection*, 40, 760–2.
Goldin, B. R. & Gorbach, S. L. (1976). *J. Natl. Cancer Inst.*, 57, 371–5.
Goldin, B. R. & Gorbach, S. (1977). *Cancer*, 40, 2421–6.
Goldin, B. R., Dwyer, J., Gorbach, S. L., Gordon, W. & Swenson, L. (1978). *Am. J. Clin. Nutr.*, 31, 136.
Goldin, B. R., Swenson, L., Dwyer, J., Soxton, M. & Gorbach, S. L. (1980). *J. Natl. Cancer. Inst.*, 64, 263.
Goodenough, E. R. & Kleyn, D. H. (1976). *J. Dairy Sci.*, 59, 601–6.
Goodhead, K., Cough, T. A., Webb, K. S., Stadhouders, J. & Elgersma, R. H. C. (1976). *Neth. Milk Dairy J.*, 30, 207–21.
Hamdan, I. Y. *et al.* (1973). *J. Dairy Sci.*, 54, 638.
Hill, I. R., Konworthy, R. & Porter, P. (1970). *Res. Vet. Sci.*, 11, 320–6.
Kavasnikov, E. I. & Sodenko, V. I. (1967). *Mikrobiol. Zh. Kyyiv.*, 29, 146, and cited in: *J. Dairy Sci. Abstr.*, 29, 3972.
Kilara, A. & Shahani, K. M. (1976). *J. Dairy Sci.*, 59, 2031–5.
Kodama, R. (1952). *J. Antibiotics*, 5, 72.
Lencner, A. A., Lencner, Ch. P., Mikelsaar, M. R., Tjuri, M. E., Toom, M. A., Väljaots, M. E., Silov, V. M., Lizko, N. N., Legenkov, V. & Roznikov, M. J. (1984). *Die Nahrung*, 28(6/7) 607–13.

Lizko, N. N., Silov, V. M. & Syryoh, G. D. (1984). *Die Nahrung.*, **28**(7/7) 599–605.
Luckey, T. D. (1979). *Am. J. Clin. Nutr.*, **32**, 109–112.
Mann, G. V. (1977). *Atherosclerosis*, **26**, 335.
Mann, G. V. & Spoerry, A. (1974). *Am. J. Clin. Nutr.*, **27**, 464.
Mayer, J. B. (1966). *Mschr. Kinderheilkunde*, **114**, 67.
Metchnikoff, E. (1908). *The Prolongation of Life.* G. Putnam's Sons, New York.
Mott, G. E., Moore, R. W., Redmond, H. E. & Reiser, R. (1973). *Lipids*, **8**, 428–31.
Neumeister, K. (1969). *Rad. Biol. Ther.*, **10**(6) 843.
Przybylowski, P. *et al.* (1978). *XXth Int. Dairy Congr. E.*, 547.
Rašić, J. L. & Kurmann, J. A. (1978). *Yoghurt: Scientific Grounds, Technology, Manufacture and Preparations.* Tech. Dairy Publ., Copenhagen.
Reddy, G. V. & Shahani, K. M. (1971). *J. Dairy Sci.*, **54**, 748.
Reddy, K. P., Shahani, K. M. & Kulkarni, S. M. (1976). *J. Dairy Sci.*, **59**, 191–5.
Rowland, I. R. & Grasso, P. (1975). *Appl. Microbiol.*, **29**, 7–12.
Shahani, K. M. & Chandani, R. C. (1979). *J. Dairy Sci.*, **62**, 1685–94.
Shahani, K. M., Vakil, J. R. & Kilara, A. (1976). *Cultured Dairy Products J.*, **11**, 14–17.
Shahani, K. M. & Ayebo, A. (1980). *Am. J. Clin. Nutr.*, **33**, 2448.
Shimada, K., Bricknell, R. S. & Finegold, S. M. (1969). *J. Infec. Dis.*, **119**, 273.
Smirnov, K. V., Syrykh, G. D., Legonkov, L. G., Goland-Ruvinova, V. I., Medkova, I. L. & Voronin, L. I. (1982). *Kosmiceskaja Biol. Aviasm. Med. USSR*, **16**(2) 23–6.
Speck, M. L. (1978). *Dev. Ind. Microbiol.*, **19**, 95.
Stadhouders, J., Cordes, M. M. & Van Schouwenburg Vanfocken, A. W. J. (1978). *Netherlands Milk and Dairy J.*, **32**, 193–203.
Tanaka, R. & Shimosaka, K. (1982). *Japanese J. Geriatr.*, **19**(6) 577–82.
Vakil, J. R. & Shahani, K. M. (1965). *Bacteriol Proc.*, 9.
White, J. W. Jr. (1975). *J. Agr. Food Chem.*, **23**, 886–91.

Chapter 4

PROPERTIES OF YOGHURT

Stanley E. Gilliland

Animal Science Department, Oklahoma State University, USA

Cultured yoghurt can be characterized as a gel-like coagulated milk product, having a smooth consistency and a pleasing tart flavour. It is available in a wide range of variations, including plain yoghurt and many fruit, vegetable and/or nut flavoured varieties. Yoghurt is also available as drink products and as frozen desserts.

The definition of yoghurt in the FAO/WHO Standards is 'the coagulated milk product obtained by lactic acid fermentation through the action of *Lactobacillus bulgaricus* and *Streptococcus thermophilus* from milk and milk products. . .'. The standards also make provisions for the use of other cultures as optional additives (Winkelmann, 1986). Traditionally, the manufacture of yoghurt is generally considered to be dependent on the fermentation of a 'milk mix' by a starter culture composed of *Lac. bulgaricus* and *Str. thermophilus*. The composition of the 'milk mix' may be varied depending on the desired formulation with regard to butterfat content, total solids, and stabilizers.

A yoghurt-like product can be manufactured by the direct acidification of milk with acidulants, such as gluconodelta lactone. When fruit or other flavouring ingredients are added, the product very closely resembles cultured yoghurt containing similar additives. However, there is no evidence to suggest that such products would provide any of the health or nutritional benefits of cultured yoghurt. Thus, they will not be further discussed.

The cultures used to manufacture yoghurt are characterized as thermophilic lactic starter cultures (Thunell & Sandine, 1985). The organisms primarily included are *Lac. bulgaricus* and *Str. thermophilus*. Both of these organisms contribute to the properties of the product during the fermentation process. The primary function of both of these cultures is to produce the lactic acid required for the formation of the coagulum. Additionally they produce certain volatile compounds, such as acetaldehyde and acetone, which contribute to the flavour of the product. The latter are produced primarily by *Lac. bulgaricus* (Gilliland, 1985a). The combination of the lactic acid and volatile components is responsible for the typical flavour and texture of cultured yoghurt.

While *Lac. bulgaricus* and *Str. thermophilus* are the primary organisms involved in the manufacture of cultured yoghurt, other starter cultures also have been utilized for producing yoghurt or similar products. One of these is *Lactobacillus acidophilus* which may be used to produce a yoghurt-like product. For some products, cells of *Lac. acidophilus* or bifidobacteria may be added to yoghurt which has been fermented by *Str. thermophilus* and *Lac. bulgaricus*. Caution should be exercised in adding cells of *Lac. acidophilus* to cultured yoghurt because of the possibility of the cells of *Lac. acidophilus* not surviving during storage of the product (Gilliland & Speck, 1977). Because of this, it would be necessary to select a strain of *Lac. acidophilus* which would survive during storage in the product. Since the potential benefits of *Lac. acidophilus* in fermented dairy products will be covered in another chapter in this book, it will not be discussed in this chapter; rather the focus will primarily be on the organisms normally utilized to manufacture traditional cultured yoghurt.

MANUFACTURE OF YOGHURT

A typical schedule for the production of yoghurt is shown in Fig. 1, and certain features are of especial relevance in the present context. In particular, it should be noted that:

(1) It is usual to raise the total solids-non-fat of the process mix above that of normal milk, so that on a weight/weight basis, yoghurt will have an elevated level of protein and minerals. It may also have a higher lactose content.
(2) The comparatively severe heat treatment changes the nature of the milk proteins (see Chapter 1), and this modification may well render the protein more accessible to digestive enzymes.
(3) Although *Str. thermophilus* and *Lac. bulgaricus* are almost universally employed to produce yoghurt, it is important that numerous strains of these species are used in the industry. Some strains of

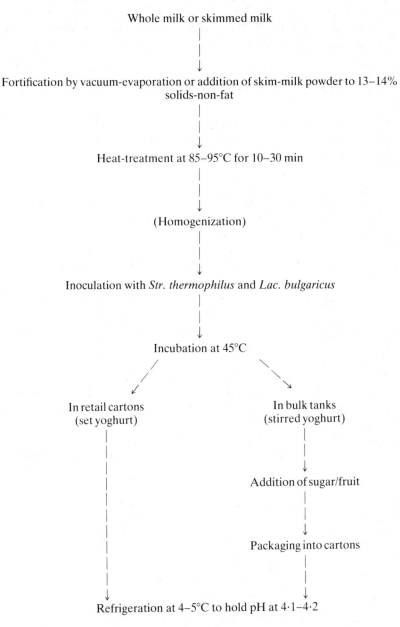

FIG. 1. The principal steps employed in the production of yoghurt.

Lac. bulgaricus, for example, are more active in proteolysis than others, and similarly while some strains produce copious amounts of extra-cellular polysaccharide, others produce little, if any)

CHANGES IN MILK RESULTING FROM FERMENTATION

The growth and action of a starter culture composed of *Lac. bulgaricus* and *Str. thermophilus* during the manufacture of yoghurt can create a number of changes in the milk. These are summarized in Table I. The most obvious change in milk resulting from the fermentation by the yoghurt starter culture is the formation of the gel-like coagulum resulting from the precipitation of casein. The yoghurt culture produces a mixture of D and L lactic acid during the fermentation (Tamime & Deeth, 1980; Gurr, 1986; Rašić, 1987). The production of lactic acid is from the fermentation of lactose in the milk, thus the lactose content is decreased during the fermentation. As much as 20–30% of the lactose present in the original milk is utilized during the fermentation (Gilliland & Kim, 1984; Rašić, 1987). Only that amount which is required for growth and acid production by the culture is used during the fermentation. The yoghurt starter cultures preferentially utilize the glucose moiety of lactose, which often results in the accumulation of free galactose in the yoghurt (Gurr, 1986; Rašić, 1987).

One report has indicated that feeding a diet composed entirely of cultured yoghurt to rats resulted in the formation of cataracts on the eyes of the rats (Breslaw & Kleyn, 1973). This was attributed to the presence of free galactose in the yoghurt. Such an effect, however, would not occur in humans for two reasons: (1) humans (except in rare cases of galactosaemia) are better able to metabolize galactose than rats; and (2) humans

TABLE I
CHANGES IN MILK RESULTING FROM FERMENTATION DURING MANUFACTURE OF YOGHURT

(1) Production of acid (primarily DL lactic)
(2) Partial utilization of lactose
(3) Improvement of digestibility of milk
(4) Increased amounts of free amino acids
(5) Alteration of vitamin content

do not consume a diet composed exclusively of yoghurt (Gurr, 1986).

Many people consider yoghurt to be more digestible than the milk from which it was manufactured. An in-vitro study involving a simulated gastric digestion system has indicated that yoghurt is more digestible than the mixture from which it was made (Breslaw & Kleyn, 1973). This increased digestibility has been related to the increase in free amino acids occurring during the fermentation of the yoghurt.

The yoghurt starter cultures are not generally considered to be very proteolytic, however, in order to grow in milk, it is likely that a certain amount of proteolytic activity is necessary (Gilliland, 1985a). *Lac. bulgaricus* seems to be more proteolytic than *Str. thermophilus* (Tamime & Deeth, 1980; Alm, 1983). During the fermentation of yoghurt, there is an increase in free amino acids, particularly valine, histadine, serine, and proline (Breslaw & Kleyn, 1973; Tamime & Deeth, 1980; Vaitheeswaram & Bhat, 1988). The proteolytic action is primarily on casein; however, action on whey proteins during fermentation of milk has been reported (Vaitheeswaram & Bhat, 1988). *Lac. bulgaricus* is more active on whey proteins than is *Str. thermophilus*. The proteolytic action on whey proteins results in partial denaturation. While the action of the starter culture during fermentation of milk to produce yoghurt results in some limited proteolytic action, the nutritional value of the proteins, apparently, is not significantly affected (Hewitt & Bancroft, 1985). Even though the proteolytic action may not greatly influence the nutritional value of yoghurt, it may contribute to the typical flavour and texture of the product.

The lactic acid bacteria generally utilize some vitamins during their growth. However, yoghurt cultures during fermentation of milk increase the amounts of some B vitamins, especially folic acid (Alm, 1983; Hewitt & Bancroft, 1985; Gurr, 1986; Rašić, 1987; Rao & Shahani, 1987). On the other hand, the amount of Vitamin B_{12} decreases during the fermentation process (Alm, 1983; Hewitt & Bancroft, 1985; Rašić, 1987). Even though there is an increase in some vitamins during yoghurt manufacture, evidence has not been presented that this results in increased nutritional benefits from yoghurt. Rao and Shahani (1987) have summarized the influence of starter culture bacteria on the vitamin content of yoghurt (Table II).

Yoghurt starter cultures exert very little action on milk fat (Tamime & Deeth, 1980). Their lipolytic activity is very limited; however, even limited amounts of lipolysis may have an important impact on the flavour of the product.

TABLE II

INFLUENCE OF FERMENTATION BY YOGHURT STARTER BACTERIA ON B-VITAMINS IN MILK[a]

Culture	Vitamin			
	Biotin	B_{12}	Folate	Niacin
Str. thermophilus	decrease	decrease	increase	decrease
Lac. bulgaricus	decrease	decrease	decrease	decrease

[a]Based on data in Rao and Shahani (1987).

The starter culture bacteria (*Lac. bulgaricus* and *Str. thermophilus*) may provide certain health and/or nutritional benefits (Table III). These will be discussed in subsequent sections of this chapter.

TABLE III

POSSIBLE HEALTH AND/OR NUTRITIONAL BENEFITS PRODUCED BY YOGHURT STARTER CULTURE BACTERIA

(1) Improve nutritional value of milk during fermentation
(2) Influence intestinal flora
(3) Provide enzymes to improve digestion
(4) Exert hypocholesterolaemic action
(5) Produce anti-tumourogenic actions

NUTRITIONAL VALUE OF YOGHURT VERSUS MILK

Using rats as animal models, feeding of yoghurt resulted in increased weight gains and increased feed efficiency compared to the milk from which it was manufactured (Hargrove & Alford, 1980; McDonough *et al.*, 1982). McDonough *et al.* (1982) reported that rats fed yoghurt grew faster than did rats fed milk (Table IV). One study has suggested that this increased nutritional value of yoghurt is due to a factor associated with a protein fraction of the product (Hargrove & Alford, 1980). Another study provided results indicating that differences in the vitamin content of yoghurt versus the milk from which it was manufactured were not responsible for the increased nutritional value (McDonough *et al.*, 1982).

Some results from this study suggested an enhanced bioavailability of minerals, in particular iron. However, rats fed yoghurt actually absorbed significantly less iron from the intestinal tract than did those animals fed non-fermented milk (Vonk et al., 1988).

TABLE IV
GROWTH OF RATS FED YOGHURT OR MILK[a]

Product	Average weight gain (g)[b]
Milk	115
Yoghurt	143

[a]From data in McDonough et al. (1982).
[b]Averages from six trials which included 12–15 rats/group.

The stimulated growth of rats fed with yoghurt is not related to the lactose content, or to the bacterial lactose content of the product. However, the presence of bacteria is important. It appears that the presence of *Str. thermophilus* is more important than that of *Lac. bulgaricus* (Hitchins et al., 1983). These findings suggest that the stimulatory factor(s) is associated with the cells of *Str. thermophilus*. As yet, no specific cellular component(s) has been characterized as stimulating animal growth.

ANTI-MICROBIAL ACTIONS PRODUCED BY YOGHURT STARTERS

Yoghurt is a converted form of milk which is much less prone to spoilage than is non-fermented milk. The low pH resulting from the production of lactic acid during the fermentation creates an undesirable environment for the growth of most spoilage micro-organisms other than yeasts and moulds. Thus, yeasts and moulds are the most likely types of spoilage organisms to cause problems in yoghurt. However, yeasts and moulds are relatively easy to control by proper sanitation during the manufacturing and processing of yoghurt. Thus, proper procedures utilized in the manufacture of yoghurt should result in the product having a relatively long shelf-life.

The role of starter cultures used in the manufacture of yoghurt in producing substances, other than lactic acid, that inhibit the growth of undesirable organisms has received much attention (Dahiya & Speck, 1968; Pulusani *et al.*, 1979; Gilliland, 1985*b*; Abdel-Bar *et al.*, 1987). There has been a considerable amount of interest in the possible application of these factors to foods, other than fermented milk products, as preservatives (Gilliland & Speck, 1975; Martin & Gilliland, 1980; Rao *et al.*, 1981).

POSSIBILITY OF INFLUENCING INTESTINAL MICROFLORA

The evidence is strong that yoghurt starter bacteria can exert antagonistic action toward undesirable micro-organisms in various foods. However, little definitive evidence is available showing the benefits of yoghurt bacteria in controlling undesirable micro-organisms in the intestinal tract. However, it is possible that some metabolite(s) produced by the starter bacteria during the manufacture of yoghurt may, following consumption of the yoghurt, exert inhibitory action toward some undesirable micro-organisms in the intestines.

Yoghurt has often been mentioned in the non-scientific literature as the source of desirable bacteria to colonize or recolonize the intestinal tract following oral antibiotic therapy. Similar reports also consider yoghurt to contribute lactobacilli that can produce benefits by growing in the intestinal tract. However, neither *Lac. bulgaricus* nor *Str. thermophilus* will colonize or grow in the intestines (Gurr, 1986; Rašić, 1987). Neither of these starter culture bacteria are bile-resistant. Since it is generally accepted that, to grow in the small intestines, micro-organisms must possess bile resistance, it is not likely that yoghurt starter cultures could play a major role by growing in the intestinal tract (Gilliland, 1979).

One study has presented data suggesting that *Lac. bulgaricus* grows throughout the intestinal tract using rats as a model (Bianchi-Salvadori *et al.*, 1984). However, germ-free animals were used in this study, and thus there was no competition from a normal intestinal flora. More importantly, the highest numbers of *Lac. bulgaricus* appeared in the stomach and the large intestines of the animals. Very few of the organisms were present in the small intestines, probably due to the low bile tolerance of this species of bacterium. It should be expected that there would be a higher concentration of bile in the small intestine. In the conventional animal, it is not likely that *Lac. bulgaricus* would have been able to compete with more bile-resistant organisms in the intestinal tract.

Conventional rats fed cultured yoghurt reportedly exhibited increased numbers of the starter bacteria throughout the intestinal tract (Hargrove

& Alford, 1978). However, in this particular study, a non-selective medium was used for enumerating the micro-organisms. Thus, confirmation of the increased numbers being due to increased numbers of the yoghurt starter cultures is questionable. Furthermore, the counts were made on intestinal segments taken from animals within a few minutes of having access to the cultured product. For this reason, the results may not provide evidence of increased numbers due to growth in the intestinal tract; rather they may merely reflect high numbers resulting from intakes of high levels of cultured yoghurt.

Lactobacillus bulgaricus has been shown to survive in the gastro-intestinal tract of both conventional and germ-free rats fed a diet of yoghurt continuously (Garvie *et al.*, 1984). However, *Str. thermophilus* survived only in germ-free rats. Neither of the organisms persisted when the yoghurt was fed along with a stock diet. Feeding of the yoghurt resulted in an alteration of the *Lactobacillus* flora from one which was predominantly heterofermentative to one which was predominantly homofermentative. The predominant organisms appearing in the homofermentative flora was *Lactobacillus salivarius*, which was not the species present in the yoghurt starter culture. Thus, it may be possible that the feeding of yoghurt can provide conditions in the intestinal tract which enhances the growth of certain lactobacilli present in the normal flora.

Even though the bacteria normally used in yoghurt starter cultures may not be able to grow in the intestinal tract, it is possible that some beneficial action in the intestinal tract may result from their presence, or the presence of specific enzymes associated with them. This is particularly true in relation to individuals who are classified as lactose maldigestors. This topic will be covered in the next section.

BENEFITS FOR LACTOSE MALDIGESTORS

The terms 'lactose intolerance' and 'lactose malabsorption' have been utilized to describe a situation in persons lacking adequate ability to digest lactose. Actually neither of these terms is correct. Lactose intolerance suggests an allergic reaction and lactose malabsorption implies that, in the normal situation, lactose is absorbed, which is not true. The inability of these individuals to adequately digest lactose is, for the most part, due to an insufficient level of lactase, in the small intestine, to hydrolyse the lactose ingested in milk products. Thus, perhaps a more appropriate term to describe this malady is 'lactose maldigestion'. The usual symptoms associated with this include cramps, flatulence, and diarrhoea. People having the problem normally avoid including milk products in their diet.

Two papers published simultaneously in two different journals have shown that yoghurt containing viable starter culture bacteria can improve lactose utilization in persons classified as lactose maldigestors (Gilliland & Kim, 1984; Kolars et al., 1984). If the yoghurt is pasteurized prior to consumption, the benefit is greatly diminished. Gilliland and Kim (1984) using the breath hydrogen test (BHT) showed that viable starter culture bacteria must be in yoghurt in order to obtain the greatest benefit in improving lactose digestion in humans (Table V). (The BHT is perhaps the most sensitive test for measuring lactose maldigestion in humans.) Pasteurization of the yoghurt after the fermentation process results in inactivation of the cultures along with the enzyme β-galactosidase (Speck & Geoffrion, 1980; Gilliland & Kim, 1984). It is this enzyme which is the important component in yoghurt for improving lactose utilization by lactose maldigestors.

A misconception often heard concerning the benefits of cultured dairy products in relation to their being beneficial for lactose maldigestors is that the lactose is completely utilized during the fermentation process. None of the starter cultures used to manufacture cultured milk products, including those used to manufacture yoghurt, will completely utilize the lactose in milk during the fermentation. For the most part, they utilize no more lactose than is necessary for their growth. If we assume milk contains approximately 4·5% lactose, that same milk fermented with a

TABLE V
EFFECT OF VIABLE CULTURES IN YOGHURT ON LACTOSE UTILIZATION BY HUMANS[a]

Product used as test dose in BHT[b]	Average ppm H_2 in breath[b]	Lactose (g/100 g)
Direct acid yoghurt	50·2	6·3
Cultured yoghurt (heated to kill starter)	22·8	4·2
Cultured yoghurt (viable cells present)	9·9	4·4

[a]Based on data of Gilliland and Kim (1984).
[b]BHT = breath hydrogen test; the lower the value (ppm H_2), the better lactose is used by the test subject; each value is an average from six subjects (the same subjects used to test each product).

yoghurt starter culture would contain approximately 3·5% lactose following the fermentation. While it is true that all of the lactose is not utilized, there is apparently enough utilized during the fermentation process to show some benefit for lactose maldigestors (Table V).

Animal studies have shown that the consumption of cultured yoghurt increases the amount of lactose hydrolysing enzyme in the intestinal tract (Goodenough & Kleyn, 1976; Garvie *et al.*, 1984). In considering the mode of action of yoghurt in improving lactose utilization in humans, it is important to remember that the culture is not expected to grow in the intestinal tract (Gurr, 1986; Rašić, 1987). Thus, there is no opportunity for lactose utilization due to growth of the bacteria. Cells of yoghurt starter cultures contain β-galactosidase as an intracellular enzyme, so it is protected during passage through the harsh environment of the stomach and is able to reach the small intestinal tract while still inside the bacterial cells. The yoghurt bacteria, when put into contact with bile, are able to hydrolyse lactose without growing (Gilliland & Kim, 1984). The bile apparently alters the permeability of the cells so that lactose can enter and be hydrolysed. When a lactose maldigestor consumes yoghurt containing viable starter bacteria, it interacts with bile in the small intestine and the lactose is able to enter the bacterial cells and be hydrolysed. This prevents the normal symptoms associated with lactose maldigestion.

Some variable results have been reported using flavoured yoghurts and frozen yoghurt (Martini *et al.*, 1987). Some of this variation may be due to heat treatments applied to the yoghurt prior to freezing in the case of the frozen product. On the other hand, it may relate to the influence of the various flavouring agents on the micro-organisms in the yoghurt. The level of β-galactosidase activity in fruit-flavoured yoghurt which is currently commercially available varies greatly (Gilliland & Kim, 1984). If yoghurt is to be consistently beneficial for the lactose maldigestor, the industry must have adequate quality control programmes to ensure that the product, whether flavoured or not, reaches the consumer having adequate levels of β-galactosidase activity.

POSSIBLE HYPOCHOLESTEROLAEMIC EFFECT OF YOGHURT

Daily consumption of large quantities of cultured yoghurt (2–4 litres/day) lowered serum cholesterol levels in human volunteers (Table VI). It has been postulated that this is due to a factor produced by, or enhanced by, the action of starter culture during the fermentation process involved in the manufacture of yoghurt. Hydroxymethyl glutarate has been suggested as the active factor (Mann, 1977). It is postulated that this factor inhibits cholesterol synthesis in the body, thus resulting in reduced serum cholesterol levels.

TABLE VI

INFLUENCE OF CONSUMING TWO LITRES OF MILK OR YOGHURT DAILY FOR 12 DAYS ON SERUM CHOLESTEROL LEVELS IN HUMANS[a]

Product	Serum cholesterol (mg/dl)		
	Before	After	Significant reduction
Whole milk[b]	196	177	No
Yoghurt made from whole milk[c]	193	175	Yes
Yoghurt made from skim milk[d]	211	150	Yes

[a]From data presented by Mann (1977).
[b]Four test subjects.
[c]Six test subjects.
[d]Five test subjects.

Rabbits fed a high cholesterol diet supplemented with yoghurt exhibited lower serum cholesterol levels than did rabbits on the non-supplemented control diet, or a diet supplemented with non-fermented milk (Thakhur & Jha, 1981). This hypocholesterolaemic action was reported to be due to calcium ions in the yoghurt; however, the contents of calcium ions in the yoghurt and in the non-fermented milk used in the experiments were the same. If the calcium ions were responsible, it is possible that they were more available for absorption from the yoghurt in the intestinal tract than from the milk. Some say minerals are better absorbed from yoghurt in the digestive system than from non-fermented milk (Rašić, 1987), while others say yoghurt does not provide a better source of minerals than milk (Gurr, 1986).

In yet another study involving human volunteers, consumption of both pasteurized and non-pasteurized yoghurt resulted in significant reductions of the serum cholesterol levels (Hepner et al., 1979). The subjects consumed 240 ml of the yoghurt three times/day during the trial. The data from this study further suggested that unaltered (unfermented) milk may contain some low level of hypocholesterolaemic activity. However, neither the component in the milk nor in the yoghurt was identified. Hydroxymethyl glutarate has been postulated to be the factor produced by the bacteria (Nair & Mann, 1977).

In a study in which milks fermented with *Lac. bulgaricus* or *Str. thermophilus* were fed to rats, no significant effect was observed on serum cholesterol levels (Pulusani & Rao, 1983). While there is still some debate as to whether or not yoghurt produces hypocholesterolaemic effects, the possibility exists that some benefit can be derived from the product in this regard. However, thus far, the studies which have shown a benefit in regard to serum cholesterol level have necessitated the consumption of very large amounts of yoghurt. Because of this, consumption of yoghurt itself may not develop into a practical means of helping control serum cholesterol. If, on the other hand, some factor produced by the yoghurt bacteria during fermentation of the milk is indeed responsible, there may be means for concentrating the active factor(s) into a usable volume for practical use. Additional research is certainly needed in order to clarify the possible hypocholesterolaemic effect of cultured yoghurt.

ANTI-TUMOUR ACTIVITY OF YOGHURT STARTER BACTERIA

Consumption of milk fermented by yoghurt starter bacteria inhibits the growth of certain types of tumours in both mice and rats (Ayebo et al., 1981; Reddy et al., 1983; Shackelford et al., 1983; Shahani et al., 1983; Takano et al., 1985). In one study, feeding of the fermented milk resulted in reduced proliferation of Ehrlich Ascites tumour cells in mice (Takano et al., 1985). Pasteurization of the yoghurt did not appreciably diminish the anti-tumour activity. The conclusion was that the anti-tumour action was associated with the cell wall of the starter micro-organisms.

Reddy et al. (1983) reported that diets supplemented with cultured yoghurt inhibited the proliferation of Ehrlich Ascites tumours in mice (Table VII). In efforts to isolate the factor in yoghurt responsible for producing anti-tumour actions, yoghurt was fractionated into the whey and curd fractions. The anti-tumour activity remained in the curd fraction. This tends to support the conclusion previously mentioned in which the anti-tumour action was associated with the cell wall of the bacteria, since the bacteria would be contained in the curd fraction.

Intra-peritoneal injection of viable cells of *Lac. bulgaricus* into mice activated macrophages (Peridgon *et al.*, 1986). Activated macrophages

TABLE VII

INHIBITION OF EHRLICH ASCITES TUMOUR PROLIFERATION IN MICE FED DIETS SUPPLEMENTED WITH YOGHURT[a]

Dietary supplement[b]	Percentage inhibition[c]	
	Series 1[d]	Series 2[d]
Yoghurt	28·0	24·0
Milk	0·0	0·0

[a]Data taken from Reddy *et al.* (1983).
[b]Supplements fed *ad libitum* along with regular mouse 'chow'.
[c]Based on comparisons with control groups (i.e. no milk or yoghurt) after 7 days on diets.
[d]Trials 1 and 2.

can play a role in suppressing the growth of tumours, thus such action may be beneficial in helping to control tumour growth. Similar results were obtained for *Lac. casei* (Peridgon *et al.*, 1986). However, when the organisms were administered through the oral route, *Lac. bulgaricus* had no effect, whereas *Lac. casei* still exhibited the effect. *Lactobacillus casei* is an organism, which can survive and grow in the small intestine because of bile resistance (Gilliland, 1979). Presumably, metabolites produced during its growth in the small intestine were absorbed and transported to the area where the macrophages were activated. On the other hand, since *Lac. bulgaricus* is not able to survive and grow in the intestinal tract, no benefit was derived from it when administered by the oral route. This suggests that lactobacilli capable of growing in the intestinal tract may be provided with greater opportunities for showing anti-tumour activities, particularly in relation to the consumption of fermented milk products.

The mechanism whereby yoghurt bacteria may produce anti-tumour action is not clear. No studies have been reported involving humans at this point to show whether or not actions similar to those observed in animal studies would be manifested. Possible anti-tumour activity produced by yoghurt starter cultures may be a very fruitful area for research in the future.

REFERENCES

Abdel-Bar, N., Harris, N. A. & Rill, R. L. (1987). *J. Food Sci.*, **52,** 411–15.
Alm, L. (1983). *Naringsforskning,* **27,** 2–8.
Ayebo, A. D., Shahani, K. M. & Dam, R. (1981). *J. Dairy Sci.,* **64,** 2318–23.
Bianchi-Salvadori, B., Camaschella, P. & Bazzigaluppi, E. (1984). *Milchwissenschaft,* **39,** 387–91.
Breslaw, E. S. & Kleyn, D. H. (1973). *J. Food Sci.,* **38,** 1016–21.
Dahiya, R. S. & Speck, M. L. (1968). *J. Dairy Sci.,* **51,** 1568–73.
Garvie, E. I., Cole, C. B., Fuller, R. & Hewitt, D. (1984). *J. Appl. Bacteriol.,* **56,** 237–45.
Gilliland, S. E. (1979). *J. Food Protect.,* **42,** 164–7.
Gilliland, S. E. (1985*a*). In *Bacterial Starter Cultures for Foods,* ed. S. E. Gilliland. CRC Press, Boca Raton, FL, pp. 41–56.
Gilliland, S. E. (1985*b*) In *Bacterial Starter Cultures for Foods,* ed. S. E. Gilliland. CRC Press, Boca Raton, FL, pp. 175–86.
Gilliland, S. E. & Kim, H. S. (1984). *J. Dairy Sci.,* **67,** 1–6.
Gilliland, S. E. & Speck, M. L. (1975). *J. Food Sci.,* **40,** 903–5.
Gilliland, S. E. & Speck, M. L. (1977). *J. Dairy Sci.,* **80,** 1394–8.
Goodenough, E. R. & Kleyn, D. H. (1976). *J. Dairy Sci.,* **59,** 601–6.
Gurr, M. I. (1986). Proceedings of the XXII International Dairy Congress (Milk the Vital Force), pp. 641–55.
Hargrove, R. E. & Alford, J. A. (1978). *J. Dairy Sci.,* **61,** 11–19.
Hargrove, R. E. & Alford, J. A. (1980). *J. Dairy Sci.,* **63,** 1065–72.
Hepner, G., Fried, R., St. Jeor, S., Fusetti, L. & Morin, R. (1979). *Am. J. Clin. Nutr.,* **32,** 19–24.
Hewitt, D. & Bancroft, H. J. (1985). *J. Dairy Res.,* **52,** 197–207.
Hitchins, A. D., McDonough, F. E., Wong, N. P. & Hargrove, R. E. (1983). *J. Food Sci.,* **48,** 1836–40.
Kolars, J. C., Levitt, M. D., Aouji, M. & Savaiano, D. A. (1984). *N. Engl. J. Med.,* **310,** 1–3.
Mann, G. V. (1977). *Atherosclerosis,* **26,** 335–40.
Martin, D. R. & Gilliland, S. E. (1980). *J. Food Protect.,* **43,** 675–8.
Martini, M. C., Smith, D. E. & Savaiano, D. A. (1987). *Am. J. Clin. Nutr.,* **46,** 636–40.
McDonough, F. E., Hitchins, A. D. & Wong, N. P. (1982). *J. Food Sci.,* **47,** 1463–5.
McDonough, F. E., Wong, N. P., Wells, P., Hitchins, A. D. & Bodwell, C. E. (1985). *Nutr. Rep. Intern.,* **31,** 1237–45.

Nair, C. R. & Mann, G. V. (1977). *Atherosclerosis*, **26**, 363–7.
Peridgon, G., Alvarez, S., DeMacias, M. E. N., Margni, R. A., Oliver, G. & De Ruiz Holgado, A. A. P. (1986). *Milchwissenschaft*, **41**, 344–7.
Pulusani, S. R. & Rao, D. R. (1983). *J. Food Sci.*, **48**, 280–1.
Pulusani, S. R., Rao, D. R. & Sunki, G. R. (1979). *J. Food Sci.*, **44**, 575–8.
Rašić, J. L. (1987). *Cult. Dairy Prod. J.*, **22**, 6–9.
Rao, D. R. & Shahani, K. M. (1987). *Cult. Dairy Prod. J.*, **22**, 6–10.
Rao, D. R., Reddy, B. M., Sunki, G. R. & Pulusani, S. R. (1981). *J. Food Quality*, **4**, 247–58.
Reddy, G. V., Friend, B. A., Shahani, K. M. & Farmer, R. E. (1983). *J. Food Protect.*, **46**, 8–11.
Robinson, R.K. (1977). *Nutrition Bulletin*, **4**, 191.
Shackelford, L. A., Rao, D. R., Chawan, C. B. & Pulusani, S. E. (1983). *Nutr. and Cancer*, **5**, 159–64.
Shahani, K. M., Friend, B. A. & Bailey, P. J. (1983). *J. Food Protect.*, **46**, 385–6.
Speck, M. L. & Geoffrion, J. W. (1980). *J. Food Protect.*, **43**, 26–8.
Takano, T., Arai, K., Murota, I., Hayakawa, K., Mizutani, T. & Mitsuoka, T. (1985). *Bifidobacteria Microflora*, **4**, 31–7.
Tamime, A. Y. & Deeth, H. C. (1980). *J. Food Protect.*, **43**, 939–77.
Tamime, A.Y. & Robinson, R.K. (1985). *Yoghurt—Science and Technology*. Pergamon Press, Oxford.
Thakur, C. P. & Jha, A. N. (1981). *Atherosclerosis*, **30**, 211–15.
Thunell, R. K. & Sandine, W. E. (1985). In *Bacterial Starter Cultures for Foods*, ed. S. E. Gilliland. CRC Press, Boca Raton, FL, pp. 127–44.
Vaitheeswaram, N. I. & Bhat, G. S. (1988). *J. Dairy Res.*, **55**, 443–8.
Vonk, A. D., Schaafsma, G., Dekker, P. R. & de Waard, H. (1988). *Neth. Milk Dairy J.*, **42**, 147–54.
Winkelmann, F. (1986). Proceedings of the XXII International Dairy Congress (Milk the Vital Force), pp. 691–702.

Chapter 5

ACIDOPHILUS PRODUCTS

ROBERT L. SELLARS

Chr. Hansen's Laboratory, Inc., Milwaukee, Wisconsin USA

When Eli Metchnikoff suggested in 1908 that man should consume milk fermented with lactobacilli to prolong life (Bibel, 1988), his proposals were met with much skepticism among his contemporaries. His theories were considered too complex by a number of scientists and they believed his theories had no basis upon which to reach these conclusions. But, as many of today's researchers in microbial adjunct-nutrition realize, his proposals do have some merit and need to be substantiated wherever possible to prove or disprove the advantages or beneficial effects of consuming products containing specific microbial organisms. Based upon the many scientific reports published in the literature, one might conclude that Metchnikoff's theories were relatively simple, understandable and straightforward. However, his theories are still viewed today by many in the medical community to be unfounded because insufficient clinical evidence exists to substantiate the claims that were made in 1908, as well as many of those that have been proposed since, particularly in the last decade.

Metchnikoff's theory was that the consumed lactobacilli would displace many of the micro-organisms normally occurring in the intestinal tract, some of which were known at that time to produce 'toxins' that were believed by several scientists to be responsible for reducing the life-span of many people. Little did the scientists of that era realize how accurate they were in their assessment for some of the 'why's' in this reduction. One of Metchnikoff's extraordinary theories, that 'the intestinal flora not only influenced the outcome of an infection by a pathogen, but it is also responsible for the insidious toxemia that hastens atrophy and aging' continues today to be challenged by several scientists in the scientific and medical communities.

However, there is much evidence which strongly suggests, if not offers good proof, that his theory does have substance. New levels of understanding and public awareness in a more health-conscious society throughout the world has led nutritionists, microbiologists and medical researchers to re-examine some of Metchnikoff's theories as they relate to the potential value of bacteriotherapy and bacterioprophylaxis of consuming microbial adjuncts.

Research with gnobotic animals in the laboratory, and domesticated animals in commercial operations, continues to document the symbiotic and associative relationships between man, animals, micro-organisms and viruses. In today's society, both man and animals are constantly being exposed to stressful situations, resulting in negative physiological changes. The effect of undue stress is often noted by a lack of appetite, greater susceptibility to microbial and viral infections and a loss of weight in man or no gain in weight, as in the case of domesticated animals. In humans, long-term research is continuing to elucidate those nutritional and microbial factors which influence the gastro-intestinal microbiota which, in turn, appears to have some connection to the psychological condition(s) exhibited by abnormal (and normal) human behavior (Meszaros et al., 1982). As our knowledge and understanding continues to expand in regards to the direct connections of our total dietary intake to both our physiological and psychological states of health, the greater appreciation scientific communities and a health conscious society will experience. Also, a greater enjoyment of life experiences will result over a longer period of time.

In the extensive review for the preparation of this manuscript, it became more evident that much of the skepticism today, as well as in Metchnikoff's time, is based upon some of the conflicting and variable results from the study of 'micronutgastrolography', e.g. micro-nutritive-gastro-intestinal effects—the study of effects of consuming products containing microbial adjuncts. There have been several proposed explanations of the variable results in this area (Johnson et al., 1987; Robinson, 1987). Among them are: (1) inadequate use of microbiological techniques; (2) poorly designed experiments and/or lack of proper controls; (3) use of microbial preparations that were below the 'minimum inoculum' at the time of consumption and/or administration (Robinson, 1987); (4) use of the wrong organism in the study, e.g. did not have the desirable strain characteristics, such as being sufficiently bile and/or acid tolerant to permit survival upon passage through the stomach and/or small intestines (Johnson et al., 1987); and, (5) were not stable to the many factors which influence survival, adhesion and implantation

(Nabukhotny et al., 1983; Prajapati et al., 1986). Unless the microbial adjuncts with the desired characteristics are consumed and/or administered at a dosage level above the 'minimum inoculum', variable results will be obtained from time to time depending upon the 'total' fluid and solid dietary intake by the animal or human subject. And, unless the total liquid and solid dietary intake is closely monitored and critically examined as to its effects on the gastro-intestinal microbiota, variable results may be noted from time to time. These subjects will be examined in greater detail in the text that follows.

AVAILABLE PRODUCTS

There are numerous products in which *Lactobacillus acidophilus* is incorporated that are available for human consumption (Prajapati et al., 1987). Some of these products are available as fermented dairy foods, which are consumed normally and regularly by some people in their daily diet. Other products are available as a liquid or in a dry form, most of which are generally consumed as a dietary adjunct while some, particularly in Russia, are consumed for their medical and publicly recognized therapeutic benefits (Solomadina, 1984; Nahaisi & Robinson, 1985; Hylmar & Pokorna, 1986; Prajapati et al., 1986).

The various *acidophilus* products may be categorized as shown in Tables I, II, III and IV. As one can see, there are a number of commercial products containing *acidophilus* that are available to consumers. Most of these are included as part of the normal daily diets of many people in Scandinavian and Eastern European countries, as well as in the USSR. Surprisingly, there are a number of *acidophilus* products that are considered to be therapeutic agents. Most of the products containing *acidophilus* which are considered to be therapeutic as well as prophylactic are often combined with other bacteria, as shown in Table III.

In reviewing the research studies using *Lac. acidophilus* alone, as well as in combination with other compatible micro-organisms for fermenting a variety of substrates (Table IV), one easily gets the impression that a number of investigators around the world have been interested in the nutritional and other beneficial attributes that can be obtained from consuming food products containing *acidophilus*.

Separate from the food products are the various *acidophilus*-containing 'health food-microbial adjuncts' which are marketed through health food stores, through distributors and/or through direct mail. These types

TABLE I
CLASSIFICATION OF *ACIDOPHILUS* PRODUCTS MILK BASED

Name	Physical type	Country of origin	Adjunct microflora in addition to acidophilus
Sweet acidophilus	Liquid	USA	
Nu-Trish a/B	Liquid	USA	Bifidobacteria species
Fermented acidophilus	Gel	Most countries	
A-38 fermented milk	Liquid	Denmark	Mesophilic lactic culture
Acidophilus yoghurt	Gel	Most countries	*Str. thermophilus* *Lac. bulgaricus*
ACO-yoghurt	Gel	Switzerland	*Str. thermophilus* *Lac. bulgaricus*
Cultura	Liquid	Denmark	*Bif. bifidum*
AB-yoghurt	Gel	Denmark	*Bif. bifidum* plus yoghurt culture
Biogarde	Liquid		*Str. thermophilus* *Bif. bifidum*
Bioghurt	Gel	FRG	*Str. thermophilus*
Bifighurt	Liquid		*Bif. bifidum*
Mil-mil E.	Gel		Yoghurt culture, *Bif. bifidum*
Miru-miru	Liquid	Japan	*Lac. casei*, *Bif. breve*
Yakault	Liquid		*Lac. casei*
Smetara	Liquid/gel	Eastern Europe	*Str. lactis* sub-sp. *diacetylactis*
Arla acidofilus	Liquid/gel	Norway	
Zdorov E. (plus malt extract)	Liquid	USSR	Kefir culture
Kefir	Liquid	Most countries	*Lac. delbrueckii* sub-sp. *bulgaricus*; *Str. thermophilus*; *Str. lactis*; *Leuconostoc* sp. *Sac. cerevisiae*
Biogarde ice-cream	Frozen	FRG	*Bif. bifidum*
Maiskii (cheese whey and skim milk)	Liquid/gel	USSR	Mesophilic lactics
Slavyanka (conc. cheese whey plus skim milk)	Liquid/gel	USSR	*Str. lactis* *Str. lactis* sub-sp. *cremoris*; *Str. lactis* sub-sp. *diacetylactis*; *Leu. mesenteroides* sub-sp. *dextranium*
Quarq	Gel	Many European countries	Mesophilic lactic culture
'Big M' (cheese whey)	Liquid	FRG	
Kurut cheese	Solid	USSR	*Str. lactis*; *Lac. delbrueckii*, sub-sp. *bulgaricus*
Butter	Gel	USSR	*Str. lactis* sub-sp. *diacetylactis*

of products are presented to the consumer in a variety of physical forms, quantities per package, new innovative packaging concepts—all with numerous claims as to total viable organisms present, guarantees of shelf-life, and the nutritional benefits one derives from consuming such

TABLE II
CLASSIFICATION OF *ACIDOPHILUS* PRODUCTS OTHER THAN MILK BASED

Name	Physical type	Country of origin	Adjunct microflora in addition to acidophilus
'Soy yo'	Gel	USA	*Str. thermophilus; Lac. delbrueckii* sub-sp. *bulgaricus*
Lupinseed	Gel	Korea	*Lac. casei; Leu. mesenteroides* sub-sp. *mesenteroides; Str. lactis; Str. thermophilus*
Matsoni	Liquid/gel	Eastern Europe	*Str. thermophilus*
Smetana	Liquid/gel		*Str. thermophilus*
Tvorog	Liquid/gel		*Str. faecalis; Lac. fermentum*
Suluguni cheese	Solid		*Str. thermophilus*
Prostokoasha	Liquid/gel		*Str. thermophilus*
Ryazhenka	Liquid/gel		*Str. thermophilus; Lac. delbrueckii* sub-sp. *lactis*
Yeast extract	Powder	UK	*Str. lactis; Pediococcus pentosaceus, Sac. rouxii, Sac. cerevisiae*
Yeast food	Powder	UK	*Saccharomyces rouxii; Lac. plantarum; Lac. fermentum; Mesophilic lactis*
Rye roburdough	Solid	Finland	*Lac. plantarum; Lac. casei*
Sourdough	Solid	FRG	*Lac. farciminis*, yeast species mixture
Recombined milk (NFDM+anhydrous milk fat)	Liquid/gel	UK	
Soy protein concentrate	Liquid/gel	Korea	Yoghurt culture
Homogenized soy milk	Liquid/gel	UK	Mesophilic lactic culture; yoghurt culture
Soy flour plus WPC	Liquid/gel	Italy	*Bif. bifidum;* yoghurt yoghurt culture

products. With the emergence of a more health-conscious society, particularly in the USA, the 'health food industry' is a multi-billion dollar a year industry and is still growing. The health-conscious consumer has become much more aware of what is important and much more knowledgeable about various products. They want to spend their resources more wisely. Therefore, the marketers of these products are requiring *acidophilus* cultures, mostly as freeze-dried powders, with much higher viable organisms per gram, and preparations which have sufficient

TABLE III
CLASSIFICATION OF THERAPEUTIC *ACIDOPHILUS* PRODUCTS

Name	Physical type	Country of origin	Adjunct microflora in addition to acidophilus
Lactinex	Freeze-dried powder	USA	*Lac. delbrueckii* sub-sp. *bulgaricus*
Solco Tricovac.	Vaccine	Czechoslovakia	
Infant formulae	Liquid	USSR	*Bif. longum; Bif. adolescentis*
Smetana (milk and sunflower seed oil)	Liquid	USSR	*Str. lactis* sub-sp. *diacetylactis*
Malyutka (vegetable oil)	Powder	USSR	
Malysh (w/o vegetable oil)			
Jojoba meal	Powder	USA	*Lac. delbrueckii* sub-sp. *bulgaricus*

stability to meet the rigid requirements imposed by their customers (the wholesalers and retailers) who are demanding noticeable performance and benefits. The primary requirement is extended shelf-life; upwards of 12 months at room temperature. Obviously, this is a stringent requirement and has been extremely difficult to meet. Very few preparations on the market qualify. Continued research has proven that the primary cause of the losses in viability, stability and performance during storage is directly related to water activity. Free moisture analyses are no longer applicable for determining the predictable shelf-life. (Sudoma, A. L., 1983, pers. comm.)

TABLE IV
USE OF *ACIDOPHILUS* TO FERMENT VARIOUS SUBSTRATES

Name	Physical type	Country of origin	Adjunct microflora in addition to acidophilus
Soybean milk plus delactosed UF milk	Light gel	France	*Bif. bifidum; Lac. delbrueckii* sub-sp. *bulgaricus; Str. thermophilus; Kluyveromyces fragilis* (yeast)
Soy protein concentrate	Liquid	Korea	
Aqueous plant seed extracts Cowpea Peanut Soybean Sorghum		USA	*Lac. helveticus; Lac. delbrueckii* sub-sp. *bulgaricus; Lac. casei; Str. thermophilus*
Skim-milk plus WPC	Gel	USA	
Milk	Liquid/gel	Finland	*Propionibacterium freudenreichii*
Extract of banana and tomato plus NFDM	Paste	India	
Banana extracts	Liquid	China	*Leu. mesenteroides; Str. lactis* sub-sp. *diacetylactis; Lac. casei*
Synthetic (water, stabilizers emulsifiers, alginates culture)	Frozen pudding semi-soft	USA	
Ginseng extract (plus skim-milk)	Liquid	Korea	*Str. thermophilus*

SURVIVAL OF LACTOBACILLUS ACIDOPHILUS IN PRODUCTS

As stated earlier, the consumers are demanding from the suppliers of *acidophilus* products noticeable and better performance from the products they buy. The marketers are demanding products with more stable cells. Therefore, culture suppliers are providing products with much higher populations of *acidophilus* as pure, single strains and/or in combination with other recognized microbial organisms. The health food-type products, mostly freeze-dried, are carefully monitored for water activity, since it is becoming generally recognized by reputable suppliers that the level of Aw is critical in maintaining maximal survival of the culture over the shelf-life required (usually 9–12 months at room temperature).

The selection of carrier in which the cultural organisms are suspended and appropriately blended is very important. Predicting the shelf-life of survival of the *acidophilus* cells in various types of *acidophilus* products has become increasingly important to suppliers of health food-type cultures, as well as those *acidophilus* cultures which are used to produce fermented dairy foods. Research has shown that for maximum survival of *acidophilus* in freeze-dried type cultures when stored at room temperature, the Aw is less than 0·250% (Chr. Hansen's Laboratory, Inc., 1984, unpublished).

It is equally important that the carriers used do not contain chemical or other types of compounds that are inhibitory to the added microorganism (*acidophilus*). A number of marketers of health foods containing microbial adjunct cultures do not recognize these two important factors, e.g. water activity and the absence of anti-microbial inhibitors in the carriers they use for blending with the culture. With the pressures of the consumers, and with the development of more sophisticated microbiological and chemical techniques of assaying, and the presence of *acidophilus* in various types of carrier/substrates, more recognition and emphasis is being paid to these two important issues.

Survival of *acidophilus* in fermented milk-based and/or in other fermented carbohydrate/protein-based products depends upon a number of important factors (Klaenhammer, 1985). Some of the important factors that *acidophilus* cultures must have before they are selected as candidates by commercial suppliers of cultures for use in fermented products as well as in freeze-dried preparations are: (1) the strains must be reasonably acid-tolerant; (2) they must be capable of developing populations in excess of one billion viable cells in the fermenter; (3) a sufficient number of cells must survive the rigors of various physical concentration techniques; (4) the cultures must exhibit minimal cellular damage during the preservation process (cryogenic freeze–thaw, freeze-drying, vacuum and/or air drying); and, (5) the final yields of cell mass at the time of sale must be economical. Obviously, there are other criteria which the selected strains must meet before they are finally selected as the organism of choice. These other criteria will be discussed later.

Maximum survival and activity is most often achieved via the cryogenic freeze–thaw process for those cultures that are used in the fermented type of *acidophilus* product; and, the cryogenic freezing of pellets prior to freeze-drying for those *acidophilus* products which are used for health food adjuncts (Klaenhammer, 1985; Gilliland & Lara, 1987; Gilliland, 1989).

The results of some research has suggested that for *acidophilus* products to have any chance of developing therapeutic and/or prophylactic benefits upon consumption, the minimum number of viable cells in a retail fermented product should be approximately one million cells per gram or milliliter of product (Robinson, 1987). In studies where *acidophilus* was used in conjunction with a mixed culture of *Str. thermophilus* and *Lac. bulgaricus* to produce *acidophilus*-yoghurt, the viable count Colony-Forming Units (CFUs) was, after 14 days storage at 5°C, four million per milliliter. These levels of *acidophilus* can be obtained routinely in this type of product by maintaining a lower acidity during yoghurt manufacture (Hull *et al.*, 1984; Sharma & Prasad, 1986; Robinson, 1987; Gilliland, 1989). When whole cells of *acidophilus* are added to 1–2% butterfat or whole pasteurized milk to make the commercial 'Sweet Acidophilus' or 'Nu-Trish' unfermented type milks, the level of surviving cells over the normal 14 day shelf-life at 5°C is one million CFUs/ml and 2·4 million CFUs/ml, respectively (Speck, 1978; Gilliland, 1989).

Survival of *acidophilus* in non-milk-based products varies considerably, depending upon the substrate, along with or in combination with other micro-organisms, final acidity of the fermented product, temperature of storage and the presence or absence of microbial inhibitors in the substrate (Kanda *et al.*, 1978; Nabukhotny & Cherevko, 1983; Hunger, 1985; Subramanian & Shankar, 1985; Schaffner & Beuchat, 1986; Sharma & Prasad, 1986; Mantere-Alhonen & Makinen, 1987; Miyamoto *et al.*, 1989). Most often higher survivors of *acidophilus* are found in milk-based products, fermented or non-fermented, than are found in non-milk-based products. The final pH and per cent acidity also affects the surviving population of *acidophilus* organisms; higher pH values (not below 4·2) allow greater survival (Robinson, 1987; Gilliland, 1989). The loss of beta-galactosidase activity within the cells is highest when the cultures are stored at higher temperatures (Gilliland & Lara, 1987). The presence of lysozyme in minute quantities affected the growth of *acidophilus* in an infant formula, 'Malyutka', (Nabukhotny *et al.*, 1983). However, it was concluded that the noticeable affecting level was just slightly below that concentration normally found in human milk. This raises some speculation about the survival of *acidophilus* when consumed by breast-fed babies. While the level of inhibitor may be sufficient to prevent maximum growth during fermentation, it is not sufficient to cause appreciable damage to *acidophilus* cells, as indicated by the isolation of reasonable numbers of *acidophilus* organisms from the faeces of the breast-fed babies.

One should recognize that there are a number of factors that exert themselves during the fermentation (growth) of the culture, during harvesting, preservation and storage, which can and do affect the survivability, activity and performance of *Lac. acidophilus* in products consumed by man or by animals. The influence of these factors depends on how the culture is used, which will often dictate the physical form the final culture product should be, to maximize survival and performance.

Over the years, cultures that are used for producing fermented *acidophilus* products are best preserved cryogenically (Gilliland & Lara, 1987; Gilliland, 1989). However, if the culture is to be offered as a health food and/or as a dietary adjunct, the preferred physical form is as a freeze-dried or lyophilized product.

Acidophilus cells that survive freeze-drying or vacuum drying are more sensitive to measurable levels of lysozyme or oxygall due to cell wall damage during freeze-drying (Brennan et al., 1986). As a result of freeze-drying stress, cell walls also become more sensitive to sodium chloride, more permeable to the transfer of toxic inhibitors which, among other things, cause damage to the cytoplasmic membrane. Scanning electron microscopy has indicated a loss in cell wall surface material, believed to be proteinous in nature. These losses can be mitigated and appreciably minimized when the fermented culture or harvested cell mass is blended with an appropriate cryoprotectant, and the pH adjusted appropriately prior to cryogenic preservation (Hylmar & Pokorna, 1986).

The survival of cultures as well as *acidophilus* during shelf-storage as freeze-dried preparations, is further enhanced by the application of selected moisture scavengers to maintain a minimum water activity level. Free moisture analysis *per se* on dried *acidophilus* products is insufficient and inaccurate for determining predictable stability. The determination of the water activity (Aw) is the most accurate method for predicting the stability of any freeze-dried cultural product (Gilliland, 1989). Analyses of a number of dried *acidophilus* health food-type products taken from commercial market shelves revealed levels of viable *acidophilus* cells ranging from 300 CFUs/g to 20 million per gram (Brennan et al., 1983). However, the total lactobacilli ranged from 10 000/g to 9 million/g. Some of these products did not contain the minimum level (one million) of *acidophilus* that has been proposed as the minimum to effect some noticeable nutritional, prophylactic and/or therapeutic benefit when consumed (Robinson, 1987; Gilliland, 1989). These products contained *acidophilus* strains that were not adequately preserved. By employing the latest biotechnology in culture production, preservation and storage, satisfactory numbers above the minimum are now possible throughout the shelf-life period, all at price values.

SURVIVAL AND IMPLANTATION IN HUMANS

In review of part of the published literature, there is evidence to demonstrate the utility of certain consumed lactic acid bacteria in preventing the growth and development of many gastro-intestinal organisms. *Lac. acidophilus* is one of the more beneficial organisms in this regard.

Survival of *acidophilus* and its potential implantation in the gastrointestinal system depends upon a number of factors, some of which were discussed in the preceding section, e.g. how the cells are produced, preserved and stored. There are also other factors which will be discussed later.

Survival basically depends upon the number of viable and healthy cells that survive passage through the stressful environment of the stomach. Since the pH of the stomach is at all times acidic, but is more so when empty, the degree of acid tolerance of the *acidophilus* cells ingested is critical to survival during this passage. Once through the stomach, the organisms are then subjected to varying concentrations of bile. Therefore, the survivors from the stomach must be tolerant to bile and reduced redox potentials if they are to survive the duodenum and the small intestines (Gilliland, 1989).

A number of reports have demonstrated the survival and implantability of *acidophilus* in humans and animals (Gilliland, 1989; Lidbeck *et al.*, 1987; Fuller *et al.*, 1986; Conway *et al.*, 1987; Pettersson *et al.*, 1983; Camschella *et al.*, 1983; Savage, 1978; Sellars, R. L., 1966, 1967, unpublished). In a clinical trial, involving 21 subjects, to study the effect of oral supplements of *acidophilus* on fecal enzyme activity, when the microbial adjunct was ingested at the rate of two million viable cells per milliliter and up to eight ounces per day over a four-week period, surviving/viable active *acidophilus* organisms (the same as those ingested) were isolated routinely from the faeces, and up to two weeks after cessation of consumption. It was concluded that the organisms not only survived passage through the gastro-intestinal system, but most likely had been implanted (Goldin & Gorbach, 1984). An independent study by other investigators who studied the contents of human jejunal aspirates found *acidophilus* survived passage through the stomach and remained viable in the proximal small bowel. The survival times varied between individuals, which was not unexpected (Luckey *et al.*, 1983).

In a related study where a human fetal intestinal epithelial cell line was used to monitor the adherence of various *acidophilus* strains, two mechanisms were demonstrated (Kleeman & Klaenhammer, 1982). One mechanism, requiring calcium in the adherence reaction, was found to be non-specific and allowed all strains to adhere. The other system, not requiring calcium, was demonstrated for only four of the 32 strain isolates studied. These were all human isolates, which suggested that implantation of *acidophilus* by adherence of the cells to the epithelial mucosa is host-specific. This has also been suggested by other investigators who found similar results not only for human intestinal epithelial cell lines, but for specific animal cell lines as well (Mautone *et al.*, 1982; Maeyrae-Maekinen *et al.*, 1983; Watkins & Miller, 1983; Fuller *et al.*, 1986).

More recently, the ability of host-specific *acidophilus* to adhere to epithelial mucosal cells was found to be directly associated to the presence of a sticky polysaccharide layer on the cell wall. Proof of this was demonstrated by the electron microscopical examination of individual cells that adhered to the tissue culture cells before they were stripped of this 'sticky' layer, but did not after the cells were treated to remove this layer (Hood & Zottola, 1987). These results were similar to those reported earlier (Savage, 1978). An assay technique to pre-screen potential *acidophilus* strains that are intended to be used in probiotic therapy for their ability to adhere to animal cell lines has been developed and used with reasonable success (Chr. Hansen's Laboratory, Inc., 1987, unpublished). This same technique has been used to pre-screen potential candidates for cultured dairy foods as well as for those used in dietary adjuncts. Further research continues to improve the accuracy and efficacy of such techniques.

From the various reports, survival and implantation of *acidophilus* strains is possible in many human systems providing the strains used are carefully selected, meet the various criteria necessary to demonstrate survivability and implantability (adherence) and demonstrate some beneficial effects. Associated with the above is the impact of the total diet, which will be discussed later; and, the total physiological and psychological state of the body at the time of ingesting *acidophilus* products and for periods thereafter. As will be discussed later, the 'total' system plays a key role in the results obtained when consuming *acidophilus* products.

FACTORS AFFECTING SURVIVAL, IMPLANTATION AND MAINTENANCE OF AN ACIDOPHILUS MICROFLORA IN HUMANS

Several factors that can influence the survival of *acidophilus* in passage through the human gastro-intestinal system have already been discussed to some extent, as have those factors related to the survival and implantation of *Lac. acidophilus*. However, there are additional factors which can and do influence not only the survival, but also the implantation of the *acidophilus* cells in one's system. There are also other forces which can often play a noticeable role in maintaining a balanced microflora in the intestinal systems.

Micro-organisms, which are candidates for use as dietary adjuncts, should have the following characteristics before they are selected and used for human consumption. As pointed out earlier, one of the most important characteristics in an organism used for this purpose is that it should be a normal inhabitant of the intestinal tract. The selected strain should have been isolated from human faeces or from an implantation of the culture within the host. It must be host specific (Gilliland, 1979; Klaenhammer, 1985; Kim, 1988). Also, as indicated earlier, the selected strain must be able to survive and grow in the small and large intestines (bile tolerant) after its survival in passage through the stomach and duodenum sections of the 'gastro' system (Gilliland, 1989).

Another important factor that contributes directly to its establishment and maintenance in the intestinal system is the ability of the strains to produce some noticeable benefits. Also, the selected organism must be produced, preserved and stored properly to maintain its viability and activity in the carrier food or substrate before consumption. Another equally important characteristic is the *acidophilus* strain used must be highly competitive in association with the microbiota normally found in the digestive system.

Survival of *acidophilus* during its passage through the stomach, as outlined earlier, is influenced directly by the methods used for the production, preservation and storage of the total culture. Associated with this is the influence of water activity. The amount of water, free and bound, within the cells as well as in the carrier, affects the steady but reduced enzymatic activity of the cells during storage. It could also affect the hydrophobic nature of the organisms. Research has shown that many *acidophilus* strains contain a surface protein layer which has been demonstrated to be hydrophobic (Bhowmik *et al.*, 1985; Johnson *et al.*, 1987). This protein layer could be removed and extracted by mixing the cells with sodium dodecyl sulfate or quaindine hydrochloride. Analysis of this protein layer revealed it to be composed of about 30% hydrophobic amino acids, even though it was shown to be resistant to several proteolytic enzymes. Once this layer was removed or destroyed, the hydrophobicity of the cells was lost, and death followed immediately. Even when employing a variety of growth conditions, these factors did not maintain the synthesis of this protein coat once it was removed. However, there are fermentation parameters and techniques for preservation prior to storage which significantly minimize the loss of this 'coat' (Chr. Hansen's Laboratory, Inc. 1989, unpublished). It is suggested that this 'coat' is part of the fibrils, which have been demonstrated through scanning electron microscopy (SEM), and which were found to be strongly evident in those strains that showed the ability to adhere to epithelial intestinal mucosal cells (Sandine *et al.*, 1972). It is not illogical to believe that the 'fibrils and the sticky layer' work in tandem to provide the mechanism whereby the *acidophilus* cells can adhere to epithelial cells, thereby, offering opportunities for implantation otherwise not available to other microbial species. The only microbial species that have been demonstrated to have this unique ability of implantation are selected species of bifidobacteria. To keep both the carbohydrate 'sticky' and the fibril protein 'coat' in optimum condition, the hydrophobicity must be controlled and regulated effectively. Some of the new biotechnological improvements in stability, achieved by maintaining the optimum Aw, do this effectively. If the Aw gets too high, the cells expend excess energy thereby reducing their life-span (reduced stability). On the other hand, if the Aw is excessively low, the phobicity of the fibrils and sticky layers change, probably become more brittle and not sufficiently active to effect adherence and eventual implantation.

It is well known that the gastro-intestinal tract of healthy, well-fed animals and humans contain micro-organisms of a diverse nature. Replicate analyses of faeces have shown that the microbial population in the intestinal tract can exceed 100 million cells per gram of content. When calculating the weight of the entire intestinal contents and the average

CFUs/g, the total microbiota in the system has been estimated to exceed 100 trillion cells. This would be at least 10-fold or more than the population of human cells in the body. Maintaining the diversity of this population in proper relationship with the important lactobacilliacae found in normal healthy intestines is important for continued good health.

Another important characteristic for *acidophilus* strains to possess, if they are to have the chance of survival and implantation in the digestive tract, is the ability to compete with the complex biota. Many lactobacilli by nature produce microbial inhibitors (bacteriocins) which are antagonistic, and often bacteriocidal and viracidal, towards many pathogens, as well as the many Gram-negative asporogenous organisms commonly found in the digestive tract (Gilliland & Speck, 1977; Klaenhammer, 1985). Some of these bacteriocins have been characterized and found to have a low molecular weight (264 kd), while others were found to be slightly higher (Kim, 1984; Sharma & Prasad, 1986). Therefore, before the culture can demonstrate any benefit, it is important that the *acidophilus* cells survive implant and grow in sufficiently high numbers by competing with the other flora present in the system. Without this ability, little to no nutritional, prophylactic or therapeutic benefits will be realized.

Is it illogical to assume that the consumption of *acidophilus* organisms in a variety of products, when consumed in sufficient numbers, have a profound influence on the diverse microflora in the intestinal tract? Since most of the microbes found in the intestinal tract occupy niches in distinctive habitats and become indigenous to those areas, it is important that any lactic culture used for fermented dairy products, or used as dietary adjuncts, have the ability to compete with the normal intestinal microflora. The presence of specific genera and species of the natural intestinal-flora, which occupy these areas in significant quantities, form together a natural 'ecological barrier' against other invasive micro-organisms, including the lactic acid bacteria. However, when viable *acidophilus* organisms with antagonistic properties are consumed, perhaps they are able to establish their own 'beachhead' and act as a protective barrier against significant pathogens which can be most invasive. When *acidophilus* is present in sufficiently high numbers, they are capable of 'translocating' across the GI lumen mucosa (Berg, 1983; Bianchi-Salvadori *et al.*, 1984, 1986). Translocation of the lactics has been demonstrated only for *acidophilus* and a specific strain of *Lac. bulgaricus*. However, translocation of other microbes, such as *E. coli, Pseudomonas, Klebsiella*, and clostridia has also been demonstrated (Berg,

1983). These studies were done in gnobiotic animals. However, it is not illogical to assume that the phenomenon also occurs in humans, as will be discussed later.

The key to effectively controlling the translocation of *acidophilus* organisms into the lymph system, spleen and liver is to daily ingest sufficiently high numbers of active *acidophilus* cells with the desirable characteristics discussed previously. Ingesting high numbers helps to moderate the microbiota in the GI tract, and they provide some protection against invading pathogens, by regulating the 'ecological' conditions and interactions among the highly diverse microflora found in the human alimentary system.

The production and elaboration of the above-described antagonistic factors to compete with the diverse micro-structure in the digestive tract is not the only factor that helps to contribute to the competitive nature of *acidophilus*. The proposed mechanisms for controlling the flora, in addition to 'bacteriocins' are: (1) lowering of the intestinal pH; (2) competing for nutrients; and, (3) production and elaboration of antitoxins (antienterotoxins) (Pollman *et al.*, 1984; Reddy *et al.*, 1983, 1988). The antimicrobial action of lactic acid and acetic acid, as produced by the *acidophilus* and bifidobacteria, on Gram-negative microbes, as well as several of the Gram-positive pathogens, is well recognized. Therefore, the ability of the *acidophilus* cells surviving through the system to the intestinal tract to hydrolyze lactose is important.

A number or reports have shown the effects of different preservation techniques on the activity of beta-galactosidase (beta-gal) (Prajapati *et al.*, 1986; Johnson *et al.*, 1987; Prajapati *et al.*, 1987). Variable beta-gal activities, not only between different preservation techniques but within the same techniques, were observed. This would suggest that perhaps the average age of the cells used in the study was not optimum, e.g. late log phase of growth (Porubcan & Sellars, 1979; Sellars, 1967). Unless satisfactory activity of beta-gal is maintained during preserved storage and during passage through the digestive tract until the cells become implanted, their maintenance in the 'niches' will not be effective. Obviously, their proteinase systems must be operating effectively for survival and maintenance within the 'niches'. It is important when consuming *acidophilus* dietary adjunct products, that one drinks at least eight ounces of milk daily. For continued maintenance, the presence of lactose is necessary (Sellars, R. L., 1967, unpublished). When consuming *acidophilus* products in a milk-based formula, sufficient lactose is available. However, one should note the quantity of milk consumed with the product to be sure that the daily consumption of fluid milk will suffice.

One's total dietary intake of solid food and liquids will directly affect the maintenance of *acidophilus* in one's system (Sellars, R. L., 1967, unpublished). A number of clinical research reports on the efficacy of *acidophilus* in producing direct, positive benefits have been variable at best. Some results were neither positive nor negative. In most reports, no concerted effort was made to control the 'total' dietary intake of the human test subjects. While in many experiments with germ-free animals, the total diet was measured and analyzed both microbiologically and chemically. In many of the gnotobiotic animal experiments where *acidophilus* was used as a dietary adjunct or for a specific purpose, most of the results have shown some positive effects and benefits. Why? It is the author's contention that the total diets of the clinical human test subjects were not rigidly controlled nor taken into account accurately when analyzing the results. In some of the experiments, negative results were obtained; however, in others, positive results were noted (Goldin & Gorbach, 1984). In those experiments where the test subjects were monitored and/or where the dietary protocol was adhered to by the test subjects, the benefits of consuming *acidophilus, per* the experimental objectives, were realized to some degree.

In other experiments where variable to negative results were obtained, one could speculate that the *acidophilus* culture(s) used in the trials were ineffective because they did not meet the requirements as stated earlier, e.g. stability, survival, activity, implantability and maintenance; or, the level of administration was below the minimal effective level; or, as in some cases, the diets of the subjects contained raw vegetables, many of which contain natural beta-glucoside type compounds which are antagonistic to lactobacilli (Fleming *et al.*, 1973). Several of the tuber vegetables contain natural microbial inhibitors which protect the vegetable species during its growth and development in soil. When consumed raw, the concentration is sufficient to retard and, in some cases, inhibit a lactic acid controlled fermentation. Raw onions, garlic and raw cucumbers, when consumed in normal quantities, affected the normal maintenance of the author's personal *acidophilus* implant. This effect was monitored carefully a number of times. When these vegetables were sufficiently cooked, the adverse effect on the implant was not observed, which was not too surprising. Nature provides numerous examples of 'natural antagonistic inhibitors' to protect the young growing plant and animals from microbial, viral and/or mycotic infections. This protection is obviously not 100%. However, these systems provide 'natural' protection, thus allowing the perpetuation of the species of normal healthy specimens.

Another factor in the total dietary intake which most likely is not taken into account when conducting clinical trials with *acidophilus* in human test subjects, at least it is never mentioned, is the impact of the liquid intake. The consumption of alcoholic beverages can directly influence effective implantation and maintenance of *acidophilus* organisms in one's GI tract (Sellars, R. L., 1970, unpublished). Different alcoholic liquids have varying compositions, qualitative and quantitative, of flavonoids, mercaptens and alkaloids, some of which are inhibitory to lactic acid bacteria; the consumption of red wine affects the implant much more than white wine. An examination of the chemical components of these types of alcoholic beverages reveals that the red types contain more of the inhibitors than white wine. The more the beverage is processed, the less of these chemicals are present. Heavy, dark beer is more inhibitory than light, premium beer. One could speculate that the effect of these beverages on *acidophilus* maintenance in the digestive tract is primarily due to the total consumption of the alcohol itself. However, under normal conditions in a healthy system, the activity of the alcohol dehydrogenase enzyme is sufficient to inactivate the alcohol before it comes in contact with the implanted *acidophilus* cells. Also, it seems likely that most of the remaining free, 'un-denatured' molecules would have been absorbed into the vascular system before reaching the implanted area of the intestines.

However, the other components in the beverage are not easily metabolized, as evidenced by an acute hangover when one imbibes to excess. Also, when one examines the various liqueurs/spirits for these 'other components' one will find that Scotch liquor contains less of these compounds than many of the other similar alcohol drinks. How many times have you heard people say they can drink Scotch all night and never have a hangover? The alcoholic content of consumption can be the same or more, but the 'morning after' effects are not the same, e.g. much milder. Therefore, the type and quantity of alcoholic beverage consumption does have negative effects on the maintenance of *acidophilus* in the digestive system.

I would not be surprised if you may be wondering and questioning the validity of some of the statements in the foregoing paragraph. To my knowledge, there is no published data or other information on the type of experiments which have demonstrated the effects as described. These statements are based on 12 months of personal in-vivo experiments in my system, as well as in-vitro experiments testing the extent of the antagonistic inhibitors on the growth and/or survival of *acidophilus* when sub-cultured in the presence of varying concentrations of such compounds.

Every week for 12 months, my feces were aseptically collected and plated for *acidophilus*, total lactobacilli, coliforms, staphylococci, total anaerobic spore-formers, total CFUs, and yeasts and molds.

Implant maintenance was clearly demonstrated for up to three weeks after discontinuing daily consumption of four to six ounces of a fermented milk culture of *acidophilus*, containing not less than one billion viable cells/ml, provided that eight ounces of milk was consumed daily (this provided the necessary lactose and protein for cellular activity, growth and development), and providing my diet did not contain excessive amounts of raw vegetables and no alcoholic beverages. Consumption of the same vegetables, but after they were cooked, had no adverse effect on *acidophilus* maintenance in my system. Trials to determine the 'minimum inoculum' to effect an implant in my system was also established during this period. Consuming four to six ounces of a fermented *acidophilus* culture with less than one million viable cells per ml was ineffective in establishing colonization. Whenever the population of *acidophilus* dropped below one million viable cells per gram of feces upon analysis, maintenance of colonization was difficult and often was observed to be lost by the next analysis, usually within seven days. Re-establishment was often easily accomplished by consuming a milk-based fermented culture as described above. Consuming freeze-dried, powdered *acidophilus* cultures dissolved in water were ineffective. However, suspending the preparations in pasteurized milk prior to consuming was effective, providing sufficient numbers of organisms were consumed, e.g. one million viable cells per milliliter in a minimum of four to six ounces. Allowing the dried culture to rehydrate for 10–15 min in the milk while refrigerated seem to work best.

Consumption of 500-mg capsules containing dried *acidophilus* with a concentration of one billion viable cells per gram was occasionally effective in establishing an implant. Consuming four capsules twice per day after eating were also often effective in maintaining colonization of *acidophilus* (Sellars, R. L., 1988, unpublished).

An effective maintenance program was not possible when consuming *acidophilus* products in tablet form. The number of viable cells per tablet is the least of all adjunct type products. An analysis of commercial tableted products on the market often revealed products with *acidophilus* concentrations from 100 to 1000 CFUs per tablet. And, most often the total count of organisms other than *acidophilus* was greater. Significant losses of viable cells in tableted products are a result of the tableting process itself. Pressures in excess of 1000 pounds per square inch are often used to effectively get tablets to form. Also, for a millisecond, the temperature in the die-press can be in excess of 100°C. The combination of pressures and temperatures can cause viable cell losses to exceed 99% (Chr. Hansen's Laboratory, Inc., 1979, unpublished).

For 25 years, personal in-vivo experiments have given valuable information from which many conclusions are valid. However, the results can only be attributed to one individual's system. One cannot say that the same or similar results might occur with others, but empirical evidence observed by others over extended periods of time (more than one year) indicates that they have experienced similar positive results (Gilliland, S. E., 1981, pers. comm.).

HEALTH PROMOTING PROPERTIES ASSOCIATED WITH THE ESTABLISHMENT OF AN *ACIDOPHILUS* MICROFLORA

The Webster's dictionary defines health as: 'A condition of being sound in body, mind and/or spirit with freedom of physical disease or pain.' The following health-promoting properties can be obtained when consuming *acidophilus* products when sufficient numbers of active viable cells are consumed. Colonization of *acidophilus* in the GI tract helps to promote growth, aids the digestion of lactose, helps to increase the rate of mineral absorption, produces and elaborates anti-microbial and anti-carcinogenic factors, stimulates the immune response system and stimulates the reduction of blood serum cholesterol.

The above attributes and benefits realized from the consumption of *acidophilus* products are based upon the fact that the strains of *acidophilus* consumed have the desired characteristics necessary for survival and implantation in the gastro-intestinal system. Also, the host is a reasonably healthy individual, and one that is not under undue physiological and/or psychological stress.

Promotes Growth

Consumption of *acidophilus* helps to promote growth by reducing the number of proteins requiring and consuming undesirable micro-organisms that are common inhabitants of the intestinal system. The availability of cultural proteinases and carbohydrases increases proteolysis of milk protein and other proteins which produce amino acids that can be more readily absorbed and more efficiently utilized metabolically. *Acidophilus* strains were observed to have true dipeptidase enzyme activities on a large number of dipeptides. Gel electrophoretic analysis of enzyme activities revealed that a number of lactobacilli, including *acidophilus*, were capable of hydrolyzing the whey proteins, lactalbumin and beta-lactoglobulin (El Soda & Desmazeaud, 1982). In-vivo animal experiments showed that when the animals consumed fermented milk-based *acidophilus* products, less nutrients were required to effect weight gains. In other experiments, *acidophilus* products were studied for their effectiveness as probiotics for swine dietary adjuncts. The results demonstrated that feed/gain ratios were significantly improved, and the animals maintained a more healthy state during their growth promotion phases (Aimutis, W. R., 1989, pers. comm.). Since the metabolism of the swine animal and the human system is most similar, is it totally incorrect to assume that better growth and a more healthy person would result when appropriate *acidophilus* products were consumed regularly? Similar results were reported for chickens grown under commercial conditions (Chr. Hansen's Laboratory Inc., 1980, unpublished data). The production of antibiotic-like substances helps to suppress undesirable bacteria, like the coliform group, and thereby promote more growth through the availability of more nutrients (Pollman *et al.*, 1984).

Aids Lactose Digestion
Lastase-deficient individuals may avoid milk and other dairy foods because of intolerance symptoms, e.g. abdominal cramping, excess flatulence, etc. While the terms lactose intolerance and lactose malabsorption have been used to describe a situation in individuals who lack the ability to adequately digest lactose; neither term is appropriately correct. Perhaps, a more appropriate term to describe this malady is 'lactose maldigestion' (see Chapter 4).

A number of experiments have been reported whereby 'lactose maldigestors' who consume fermented *acidophilus* (and yoghurt) products had far fewer digestive problems. During cultural fermentation the level of lactose is significantly reduced. Also, when *acidophilus* was colonized in the intestinal system, lactose utilization was improved significantly as measured by the breath hydrogen assay (Gilliland & Kim, 1984; Kolars *et al.*, 1984). However, in studies with the non-fermented *acidophilus* type products, lactose digestion was not enhanced appreciably upon immediate consumption (McDonough *et al.*, 1985); although, if appreciably colonized, noticeable improvement in utilization or absorption could result. When fermented *acidophilus* products that contain viable *acidophilus* organisms are consumed, reaction with the bile increases absorption of lactose into the bacterial cells where it can be metabolized more rapidly (Gilliland & Kim, 1984). This reaction phenomenon helps to explain why non-heated fermented products can be more easily consumed by 'lactose maldigestors'.

Increases Mineral Absorption
There are a number of reports which suggest that when fermented dairy products containing lactobacilli are consumed, the bioavailability for mineral absorption is increased. The presence of minute quantities of lactic acid influences the rate of absorbed minerals. It was suggested that one of the reasons for increased growth in animals fed fermented products was the increased bioavailability of vitamins and minerals, which are absorbed more readily and thereby enhance metabolism for improved growth efficiency (McDonough *et al.*, 1985). Other experiments involving human volunteers indicated that there were two basic categorical factors that may be operational. The absorption of minerals may be largely due to 'food factors' and/or 'physiological factors' (Allen, 1984). The consumption of fiber, oxalate and alcohol from food and drink are 'food factors', while vitamin deficiency, menopause and old age are the principal physiological factors which reduce absorption. On the other hand, it was found that lactose, lactic acid, vitamin D, calcium deficiency, pregnancy and lactation all increased absorption of minerals and vitamins. Interestingly, it has been suggested that a low intake of calcium stimulates the secretion of the parathyroid hormone, which in turn increases the synthesis of the active form of vitamin D and, subsequently improves calcium absorption. This alternate mechanism helps to insure that our bodies have sufficient calcium available for physiological enzymatic processes.

Contains Antimicrobial Factors
There are a number of reports in the literature which offer ample evidence that demonstrates the utility of certain lactic acid bacteria which retard and/or inhibit the growth of many other micro-organisms (Silles & Hilton, 1987). *Lac. acidophilus* is one of the primary organisms in this group which exhibit this phenomenon. The mechanisms of the antibiosis associated with lactic cultures are still under discussion, and these mechanisms still remain to be documented *in vivo*, conclusively. However, a number of investigators have reported that strains of *Lac. acidophilus* do elaborate inhibitory compounds, often classified as

'bacteriocins, which are very toxic to several genera (Sandine *et al.*, 1972; Pollman *et al.*, 1984; Bianchi-Salvadori, 1986; Fuller, 1986; Cole *et al.*, 1987; De Simone *et al.*, 1987; Kim, 1988). Many of the microgeneric groups contain pathogens that are often found in the human gastrointestinal system, particularly when the host body is infected or under abnormal physical and/or psychological stress (Keating, 1985).

The antibiosis has been attributed by some investigators to the metabolic by-products of lactic acid bacteria, e.g. hydrogen peroxide, organic acids (acetic, gloxylic, malonic, alpha-ketoglutaric acids).

Several possible mechanisms have been suggested to explain some of the noted protective effects obtained by colonization with *acidophilus*: lowering of intestinal pH; adhesion to the intestinal mucosa preventing colonization by pathogens; competition for nutrients; production of antimicrobial substances; and the production of antitoxins (anti-enterotoxins). Some of the strains produce hydrogen peroxide which activates the lactoperoxidase system, which is one of several natural anti-microbial systems (Bianchi-Salvadori, 1986). Some of the anti-microbial compounds produced by *acidophilus* have been labeled as: lactocidin, acidophilin, lactacin B and acidolin. Several of these 'bacteriocins' are low molecular weight compounds, some of which are as yet unidentified or characterized. However, some of these compounds have been identified and characterized (Barefoot & Klaenhammer, 1983). It could be that these compounds are the principal agents in affording protection to the host, either by direct action on the non-lactic micro-organism itself, or by somehow stimulating the host's immunological response system to react against invading pathogens, foreign proteins and/or viruses.

Contains Anti-carcinogenic Factors
When considering the health promoting properties associated with the establishment of an *acidophilus* microflora in one's digestive system, it is important to recognize the microbial inter-relationships that occur in one's digestive system. The occurrence of microbial interferences and interactions with the host, both of which can have profound influence in maintaining the ecological balance in the digestive system, must also be considered when evaluating the effects of *acidophilus* consumption and implantability. There are symbiotic, as well as dysbiotic reactions, some of which are beneficial while others are not. When the total microbial population in the GI tract is as diverse and awesome in numbers and complexity as it is, is it any wonder that we only feel good as long as the ultimate 'ultra-filter' (the intestinal tract) permits the flow of nutrients,

chemicals and microbial life across the lumen mucosa.

Some of the known toxic chemicals produced in the intestines are fecal enzymes that have been shown to contribute to the development of colon cancer (Goldin & Gorbach, 1984). By consuming *acidophilus* and associated products, the formation of potential carcinogens is effectively reduced (Friend & Shahani, 1984; Kim, 1988). Epidemiological evidence and some dietary studies have shown that consumption of *acidophilus* and/or *acidophilus* in combination with selected *Bifidobacterium* species in sufficient numbers, does reduce the incidence of colon cancer. For example: Finland has a high per-capita fat consumption rate and one of the lowest incidences of colon cancer.

Fermented dairy products are common components in the Finnish diet. Perhaps, the intestinal microflora is moderated and controlled by the high populations of lactobacilli, thus reducing the presence of, and the formation of, carcinogenic compounds (Goldin & Gorbach, 1984). The administration of *acidophilus* to rats which were also fed nitro, azo and/or amino-glucoronide derivatives, showed that the intestinal *acidophilus* flora has the ability to convert exogenously administered aromatic, nitro and azo compounds and an amino-glucoronide compound to free amines. The rate of this conversion process was directly affected by diet and the consumption of minimum levels of *acidophilus* cells.

A well-controlled clinical trial, involving 12 human volunteers in two groups of six, were studied in a switchover experimental design with two treatments—milk and unfermented *acidophilus* milk. Ingestion of either milk or unfermented *acidophilus* milk did not have any significant effect on the total aerobic counts. However, the consumption of *acidophilus* milk resulted in a decrease of coliforms. A significant increase in *Lactobacillus* counts were observed when their diets were supplemented with the unfermented *acidophilus* milk. The high lactobacilli counts were maintained for a period of time even after the *acidophilus* milk was discontinued. The ingestion of *acidophilus* milk instead of plain milk resulted in a reduced activity of the fecal beta-glucosidase and beta-glucuronidase enzymes which are known to catalyze the conversion of procarcinogens into carcinogens (Ayebo et al., 1980). Further in-vivo experiments are needed to further examine these activities and document the importance of *acidophilus* in this regard.

Does *acidophilus* have any effect on the survival of selected viruses? Because of obvious reasons, only in-vitro experiments have been conducted on the possible inhibition of viruses by *acidophilus*. Slurries of fermentable food waste containing *acidophilus* were inoculated with a series of selected viruses (pseudorabies, newcastle disease, infectious canine hepatitis, avian infectious bronchitis, measles, vesicular stomatitis, and porcine picornavirus (Wooley et al., 1981)). Samples of each were incubated at 5°C, 10°C, 20°C and 30°C for 96 h. The newcastle disease virus and the infectious canine hepatitis virus survived the entire period at all temperatures. The porcine picornavirus was inactivated at 30°C after 72 h, but survived for the entire test period at the other temperatures. The pseudorabies virus was inactivated at 20°C and 30°C within 24 h, but survived for 48 h at 10°C and 96 h at 5°C. Avian infectious bronchitis virus was inactivated at 20°C and 30°C within 24 h. However, it survived 72 h at 5°C and 10°C. The measles and vesicular stomatitis viruses were rapidly inactivated at all temperatures. This raises the speculation that perhaps fermented *acidophilus* might be even more detrimental to those viruses which survived the tests.

A number of reports have been published that have shown that *acidophilus*-based products have therapeutic effects on the treatment of *Trichomoniasis vaginalis* (Litschgi et al., 1980). In one multi-center study of the treatment of 444 women with *Trichomoniasis vaginalis*, a vaccine (solcotrichovac) product containing 7×10^9 inactivated *acidophilus* cells was given at a dosage rate of 0·5 ml per dose, one dose per day for one week. One year from the first vaccination, 427 patients were re-examined; 92·5% of those vaccinated were found to be cured of any clinical symptoms. While this is a remarkable statistic, no mention of controlling the total diet was made. Therefore, the results can be questioned. However, out of this many subjects, the chances of all having the same diet control would be very slim.

Stimulates Immunological Response System
In more recent years, the interactions between dietary lactobacilli and the immuno-competence has been studied both *in vitro*, as well as *in vivo*. A number of these studies have shown that lactobacilli, especially *acidophilus*, are not only an integral part of the host's gastro-intestinal microecology, but they play an important role and function in the host's immuno-protective system by increasing specific and non-specific immune-mechanisms (De Simone et al., 1987; Perdigon et al., 1986; Bourlioux, 1986). These investigators observed stimulation and activity by the macrophages and lymphocytes. An increase in suppressor cell production was noted, as was an increase in the production of gamma-interferon. Interferon production also increased T-cell activity. This increased production and activity was the highest when *acidophilus* cells

were observed to 'translocate' more frequently. Colonization and implanting of *acidophilus* cells always stimulated the immune system and increased the host's resistance to infections.

While some studies have reported on the stimulation of the immune system for the suppression of carcinogens, insufficient clinical evidence using human test subjects is available which conclusively shows a direct connection to the consumption of *acidophilus* products. The variable results were most likely due to an advanced state of the carcinoma and/or the subject was under chemotherapy or radiation (Dahl, P., 1988, pers. comm.), however, the in-vitro evidence *strongly* suggests that there is a direct connection (Keating, 1985). Others have suggested that the positive results obtained by feeding animals, which had implanted ascite interperitoneal tumors, with *acidophilus*, was due to the stimulation of the immune properties of the host by *acidophilus* (Friend & Shahani, 1984). Although the evidence is highly suggestive that *acidophilus* and other lactic acid bacteria used to manufacture fermented dairy products reduce the risk of colon cancer, insufficient clinical evidence is available to substantiate any medical claims. Because human lives would be at a high risk if they were used to verify studies of this nature, it is highly unlikely that *bona fide* clinical studies will ever be conducted with human subjects. There are reports, however, that have indicated the disappearance of colon carcinomas during prolonged periods of *acidophilus* therapy (Sellars, R. L., 1965, 1989, pers. comm.).

Too often in the past, when human subjects with colon cancer underwent *acidophilus* therapy, the results have been less than positive. It has been suggested that the patient's carcinomas were advanced beyond any possibility of inhibition by the immune system. Often, the patients were undergoing chemotherapy or irradiation treatments, which prevent colonization by the culture. Another theory for the lack of positive response could be that the advanced condition of the cancer had significantly compromised the host's immune system (Sellars, R. L., 1987, pers. comm.).

Fortunately, in my own personal case, the carcinomas discovered in my lower colon during a routine sigmoidoscopy exam were in the early stages of development (Sellars, R. L., 1964, unpublished data). Upon the advice of the attending proctologist, *acidophilus* therapy was prescribed. In the beginning, the culture consumed was in the form of *acidophilus* capsules that contained freeze-dried preparations of *Lac. acidophilus* of approximately 1×10^4 CFUs per gram. The capsules contained approximately 500 mg of freeze-dried culture. Two capsules were consumed after eating both morning and evening meals. An analysis of the caps for the presence of total CFUs, coliforms, anaerobes, yeasts and molds, often showed them to be more than the *acidophilus* counts. All capsules were consumed with milk.

After six months of therapy, there was no evidence of any reduction in size or appearance. Proctoscopic examinations were performed bi-monthly. While monitoring my diet, liquid intake and general well-being, no changes or difference from the norm were observed.

A decision was made to switch to a milk fermented culture whereby dosages could be significantly higher (minimum 1×10^9 CFUs/ml). Also, extra benefits might be derived from consuming a fermented culture in that the nutrients (amino acids, vitamins, etc.), and the antimicrobial compounds elaborated by the cells during its growth might be helpful. During the second six months of therapy, the total diet and liquid intake was monitored, but no attempt was made to control it. Close observations of general well-being were noted and recorded. Monthly proctoscopic examinations were also continued.

The *acidophilus* culture that was consumed in this study was fermented in 100 ml of a 10% TS reconstituted skim-milk, heat-treated 95°C for 30 min, cooled to 37°C, inoculated with a 1% mother culture, incubated at 37°C for 8–9 h and then refrigerated. For the first month, 100 ml of culture was consumed daily. This concentration proved to be too much. A slight diarrheal condition developed (Sellars, R. L., 1965, unpublished data).

The second month was spent in determining the optimum intake level. This was found to be 200–300 ml of a freshly fermented *acidophilus* culture per week, depending upon the total diet. A significant difference in the consistency, color and size of stools were quickly noted during the early stages of this second period of therapeutic treatment. Since constipation problems were routine, a positive change was noted very quickly. The urge to defecate was much stronger, and the periods and ease of elimination was significantly reduced. All these positive changes resulted in less hemorrhoidal tissue.

Continued proctoscopic examinations revealed the two malignant polyps to be gradually shrinking in size by four weeks into the second six-month period. By the end of the second six-month *acidophilus* therapy period, the cancerous polyps had completely disappeared with no evidence of their existence. A close examination of the diet and liquid intake revealed nothing that would indicate that any of the solid foods and/or liquids would be likely to have caused this phenomenon. No significant changes were noted either in the general physiological and/or psychological conditions. Unfortunately, blood samples were not analyzed chemically to follow the progress of immune factors. In 1965, some of the assays for such were not available; confirmation of malignance was done at the state medical laboratory.

Subsequent to this period, my mother had a colostomy. When comparing the location of her carcinoma and my malignant polyps via our medical records, they were determined to have initiated in the same general area, e.g. approximately 20–25 cm up from the rectum in the descending colon (Sellars, R. L., 1978, pers. comm.). Unfortunately, when her problem was diagnosed the advanced state of cancerous growth was such that *acidophilus* therapy was precluded. It was discovered during surgery that her growth had started to penetrate the outer colon wall. Fortunately, she recovered completely with no recurring problems, and has been living well for 12 years since her operation. However, she consumes capsules containing freeze-dried *acidophilus* regularly, one per day. This provides positive benefits to conclude successfully, with ease, her daily irrigation. Discontinued consumption delays the process because the feces are less soft. Her general condition, otherwise, is normal for an 83-year-old grandmother.

Details of other significant benefits that have been received by me personally are too much to be included here. Suffice to say, however, over the past 25 years of empirical observation (empirical, because no satisfactory controls comparable physiologically were available), significant events have occurred as a result of *acidophilus* therapy as described above. In my opinion, they are significant because I certainly believe I know more about my own bodily functions and reactions than anyone else. Through awareness and acute observations during personal studies, associations of 'cause and effects' were eventually recognized after replicate trials had established conditions and reactions not experienced previously. Many of these experiences and results are not viewed in medical circles with much validity, since they are not part of a long-term clinical study. Yet, it does seem that scientific observations over a 20-year span, regardless of lack of duplicate controls, should account for something. Particularly, when personal observations by others have experienced similar reactions. There is no evidence that the diets in some of the reported clinical studies were no more controlled than the diets of people who consume *acidophilus* on a daily, or other routine basis, outside of a clinical study. Difference can quite often be in the design of the study, its implementation and the interpretation of the results. Variable results are often attributed in retrospect to improperly designed experiments. Sometimes, fault may be in the execution, but frequently the cause for less than expected positive results is due to the use of the wrong dietary microbial adjunct and/or the population is below that necessary as a minimum dose for achieving noticeable results.

The competition among the intestinal biota is strong; however, sending a squad of soldiers to fight an army does not seem rational, unless the small group has a tremendous advantage in order to establish a strong beachhead and hold it through collective inducement of defensive help outside of their own perimeter. Consuming proper strains of *acidophilus* in sufficient numbers on a regular basis to maintain their colony (after all, every living system needs new blood once in a while), and maintaining a compatible total diet, defenses against known and perhaps some of the unknown microbials and viruses, yet unidentified, will occur in normal, physiologically and psychologically healthy individuals.

There are some speculations that for *acidophilus* therapy to achieve minimum success, desirable 'translocation' is important. Translocation is defined as the process whereby the micro-organism(s) and/or viruses cross the gastro-muco-lumen, and finding residence in the lymphatic system, the liver, spleen and in fetuses of pregnant women. Those reported translocation experiments have been *in vitro* in gnobiotic mice or rats (Bianchi-Salvadori, 1987; Berg, 1983; Doyle, M., 1989, unpublished). These experiments are highly significant in that these results are some of the first experimental evidence in animals where the micro-gastro-ecology is so rigidly controlled that lactobacilli, especially *acidophilus*, as well as other organisms, including pathogens, are capable of 'translocating' across the lumen mucosa. Translocation of not only micro-organisms, but small molecular weight compounds, such as bacteriocins like those produced by specific lactobacilli, could help the host keep its immunological response mechanisms energized and maintained in a healthy state.

Some of the more recent studies on immuno-regulation by the intestinal flora have shown that lactobacilli (*Lac. acidophilus*) and bifidobacteria (*Bif. bifidum*) influence the production of gamma-interferon against pathogenic bacteria. Gamma-interferon levels vary among individuals, probably due to the sensitization level in each individual as caused by the introduced pathogen. The presence of *acidophilus* did not directly stimulate interferon production, but did have a regulating effect on the release of lymphokine, thus modifying the antibody-dependent cytotoxicity against specific pathogens, e.g. *Salmonella typhimurium* (De Simone *et* al., 1987). These results might help to explain some of the reactions received in earlier experiments where human *Salmonella*-infected carriers were fed *acidophilus* milk *ad libitum* continuously until three to five negative feces samples were obtained. The consumption of 500 ml of *acidophilus* milk shortened the duration of the *Salmonella* carrier state. (Luckey *et al.*, 1983). In similar experiments involving 132 *Salmonella* carriers, the duration of the carrier state was shortened when they consumed a minimum of 500 ml of *acidophilus* milk (Alm, 1983). There does seem to be reasonable evidence that *acidophilus* either increased the immune responses that destroyed the pathogen, or the *acidophilus* antimicrobial substance reacted directly in preventing further survival of the infective organism.

Stimulates Reduction of Blood Serum Cholesterol

The results of reports on the effect of microbial dietary adjuncts in respect to possible reduction of blood serum cholesterol are inconsistent in their effects. Even though there are a number of reports which have established parameters and produced positive data that demonstrates the influence of consuming *acidophilus* milk on the reduction of cholesterol, there still remains much controversy over this subject. Most of the in-vivo studies on the effects of lactic bacterial ingestion via *acidophilus* milk products have been on mice, rats and swine. While the results obtained from the mice and rat studies have been questioned, perhaps the results from the swine experiments offer better evidence because the metabolism of swine is more similar to the human system.

Host-specific strains of *acidophilus*, isolated from pigs when shown to grow and assimilate cholesterol well in a laboratory medium, were fed to pigs. The results showed that the strains, when produced under anaerobic conditions in the presence of bile and fed to pigs, significantly reduced blood serum cholesterol levels in pigs fed a high cholesterol diet. The in-vivo assay may be used to pre-screen strains of *acidophilus* for their ability to either assimilate or directly act on the cholesterol (Gilliland *et al.*, 1985). Similar results were reported more recently. The reduction of serum cholesterol and low density lipoproteins was significant ($P < 0.01$ and < 0.08), respectively. There was no significant difference in the reduction of the serum tryglycerides or in the high density lipoproteins (> 0.23 and > 0.11), respectively.

While many of the experimental results from animals have been positive, the results from human studies remain less clearly defined. Further clinical studies on humans with a well-designed experimental parameter and with a defined and controlled 'total diet' are necessary to establish more definitely the direct or indirect influence of *acidophilus* dietary adjunct consumption on the reduction of blood serum cholesterol.

A number of positive health-promoting properties associated with the establishment of an *acidophilus* microflora in one's gastro-intestinal system have been discussed. Many more experiments could have been discussed and cited. Much research has been conducted on the attributes and potential benefits of consuming *acidophilus* products. More research is needed to elucidate and eventually prove the direct mechanisms involved. Hopefully, this will be forthcoming.

OVERALL CONCLUSIONS CONCERNING THE THERAPEUTIC ACTIVITY OF *LACTOBACILLUS ACIDOPHILUS*

Future translocation experiments may help to further substantiate the importance of these microbial dietary adjuncts for maintaining good health. Hopefully, the direct mechanisms will be elucidated. However, it is important that the total diet(s) used in both experimental animals, as well as humans be monitored closely, especially for the presence of any 'natural microbial inhibitors' indigenous to many foods themselves. Also, where possible, the total and specific microflora present in the food and nutrients provided to the experimental subjects be monitored both qualitatively and quantitatively. Having such information will help investigators to evaluate their results more accurately. While a germ-free environment in real life is not possible nor practical, it is important to study the effects of such dietary adjuncts in an atmosphere free from other organisms that could unduly influence or compromise the results.

The intestinal microbiota found in most humans does have a profound influence on one's immunological state, health and well-being. Viable indigenous bacteria, viruses and other proteins, good and bad, when present in sufficient quantities can translocate. I believe the degree of this phenomenon dictates the extent of the immune responses, and thus our state of health.

Most of the important benefits that are derived from ingesting *acidophilus* products have been discussed in some detail. Some of the more conclusive experimental evidence has been reviewed. One could include much more supportive information on each of the specific points and factors of importance. This is not possible in this chapter.

The reports in the literature concerning the attributes of products containing lactobacilli and bifidobacteria species which are used as dietary adjuncts are numerous. A number of published reports on the effects of *acidophilus* when tested under a variety of conditions, both *in vivo* as well as *in vitro*, have shown positive results, both therapeutically and prophylactically. While some mechanisms that demonstrate the therapeutic effects are known, most are probably not. Being able to prove the therapeutic effects and mechanisms involved during *acidophilus* therapy in humans is most difficult. However, in recent years as more accurate assays and sophisticated equipment become available at reasonable costs, the experimental investigations using microbial adjuncts are reporting results with confident interpretations. This trend should continue, but obtaining the desired accuracy when using human test subjects under clinically controlled conditions will always continue to be most difficult. The more diverse and complex the organism (human body), the more complicated it becomes to achieve results with reasonable accuracy.

Investigators need to become more creative and imaginative in their approaches to their studies of microbial dietary adjuncts. No longer can one ignore the total dietary factors in their assessments and interpretation of the results. No longer can the use of cultures that do not meet the necessary criteria be used in such experiments. More definitive guidelines for experimental designs are needed.

Definitive specifications and characteristics that dietary adjunct cultures must meet before they can qualify for use in the types of experiments discussed throughout this chapter need to be developed and generally agreed on by the scientific community involved in these types of studies. Suppliers of lactic cultures need to do a better job in classifying their commercial preparations; and, accurately determine the desirable characteristics necessary to meet consumer needs and demands.

In reviewing the published literature on the potential therapeutic activity of *acidophilus* in humans, there are a number of factors that must be considered before positive results are possible. First, the particular strain of *acidophilus* used experimentally or taken as a dietary supplement must meet certain criteria: (1) the strain must have been a normal inhabitant of the human intestinal tract, e.g. must be host specific; (2) the organisms must be sufficiently acid and bile tolerant to allow survival through the upper GI tract; (3) the cells must be competitive and able to grow and survive in the intestinal system under stressful conditions; (4) a majority of the cells must have the ability to adhere and implant in the intestinal tract before colonization; (5) the selected culture must demonstrate some beneficial effect, e.g. produce anti-microbial and anti-carcinogenic factors, stimulate the immune system, or have anti-cholesterolemic properties, and/or anti-viral activity.

Second, the organisms must be consumed and/or administered in sufficient numbers (above the minimum inoculum) to have any possible positive benefits.

Third, the culture must be produced under the most stringent GMP conditions, be preserved with maximum viability and activity, and have the stability to meet market demands.

Fourth, organisms selected for therapeutic and/or dietary adjuncts must meet all the above at a 'price value', at a cost that consumers can afford. Unless the products are consumed for whatever the reasons, no benefits will be realized.

The data, overall, shows positive therapeutic benefits when the criteria and use of *acidophilus* meet those requirements and criteria presented previously. The consumers need to be aware that their total diet will directly affect the degree of benefits obtained from ingesting *acidophilus* products. The intestinal microbiota found in most humans does have a profound influence on the immunological state, the immune response, overall general health and well-being. Viable indigenous bacteria, viruses and other proteins, good and bad, when present in sufficient numbers, can translocate. The degree or extent of the translocation phenomenon directly dictates the degree of the immuno-responses, and thus our state of health.

For the dietary microbial organisms to demonstrate some positive benefit, the consumer must realize that positive reactions are rarely achieved overnight. Sufficient time is required in order for the organisms to get established in one's system. One should not be impatient. Another important factor, and perhaps the most important of all, is the consumer needs to be aware of what they are really doing, what they should be looking for; and be cognizant of any subtle change in their bodies, physiological and/or psychological functions. Once different states of normality or a bench mark for control comparison is established for various activities within the human body, one realizes a change has taken place and gradually observes differences in one's general state of health in comparison to one's overall condition previously. Once one understands the signals, one can then begin self-experimentation to find the best combination of factors to maximize the benefits of consuming *acidophilus* products. When the positive observations are confirmed repetitively, the consumer becomes convinced that *acidophilus* consumption is important to maintain good health.

POSSIBLE ADVERSE EFFECTS

Wherever there is a positive, there is a negative also. While there are not many negatives to the consumption of lactobacilli and/or bifido microbial adjuncts, there are a few. Anyone sensitive to milk protein should be aware of the risks involved with the consumption of *acidophilus* products that contain milk or milk derivatives. Otherwise, there have been no reported adverse effects from the consumption of *acidophilus* products. One of the most common negative effects is the over-anticipation by the consumer that dramatic benefits should be forthcoming in a short period of time; and, that the results they are looking for will continue for long periods of time without much effort. What the consumer perceives should happen is not always practical, nor possible. Education of the consumer is important and necessary if the future of *acidophilus* therapy is to be realized to its maximum. Honesty and truth in advertising and marketing *acidophilus* products is becoming critical to achieving the success one can have and should have when consuming *Lac. acidophilus* in some form or another. How well this is accomplished remains to be seen. However, responsible scientists and business people need to be of one mind in their co-operative approach to provide better nutrition and health to all members of the society. If this responsibility is accepted and implemented to the best of everyone's ability, the future of maintaining a healthy society will be better realized.

DRINK UP!

REFERENCES

Aimutis, W. R. (1989). Probiotics and Nutrition. (Personal Communication). Chr. Hansen's Laboratory, Inc. Milwaukee, WI.

Allen, L. H. (1984). *Nutr. Press*, **47**(1), 1.

Alm, L. (1983). *Progress in Food and Nutrition Sci.*, **7**, 13–17.

Ayebo, A. D., Angelo, I. A. & Shahani, K. M. (1980). *Milchwissenschaft*, **35**, 730–3.

Barefoot, S. F. & Klaenhammer, T. R. (1983). *Appl. Environ. Microbiol.*, **45**, 1808–15.

Berg, R. D. (1983). *Curr. Microbiol.*, **8**, 285–92.

Bhowmik, R., Johnson, M. C. & Ray, B. (1985). *Int. J. Food Microbiol.*, **2**, 311–21.

Bianchi-Salvadori, B. (1986). *Int. J. Immunotherapy, Suppl. II*, 9–19.

Bianchi-Salvadori, B. (1987). IDF F-Doc. 136.

Bianchi-Salvadori, B., Camashella, P. & Bazzigaluppi, E. (1984). *Milchwiss.*, **39**, 387–91.

Bibel, D. J. (1988). *ASM News*, **54**, 661–5.
Bourlioux, P. (1986). *Cah. Nutr. Diet*, **21**, 204.
Brennan, M., Wanismal, B. & Ray, B. (1983). *J. Food Protect.*, **46**, 887–92.
Brennan, M., Wanismal, B., Johnson, M. C. & Ray, B. (1986). *J. Food Protec.*, **49**, 47–53.
Camaschella, P., Bianchi-Salvadori, B. & Laria, L. (1983). *Ann. Microbiol.*, **33**, 109–19.
Cole, C. B., Fuller, R. & Newport, M. J. (1987). *Food Microbiol.*, **4**, 83–5.
Conway, P. L., Gorbach, S. L. & Goldin, B. R. (1987). *J. Dairy Sci.*, **70**, 1–12.
De Simone, C., Ferrazzi, M., Di Seri, M., Mongio, F., Baldinelli, L. & Di Fabio, S. (1987). *Int. J. Immunother.*, **3**, 151–8.
El Soda, M. & Desmazeaud, M. J. (1982). *Can. J. Microbiol.*, **28**, 1181–8.
Fleming, H. P., Walters, D. & Etchells, J. L. (1973). *Applied Micro.*, **26**(5), 777.
Friend, B. A. & Shahani, K. M. (1984). *J. Food Protect.*, **47**, 717–23.
Fuller, R. (1986). *J. Appl. Bact.*, (Symp. Suppl.), 1S–7S.
Fuller, R., Cole, C. B., Newport, M. J. & Ratcliffe, B. (1986). *Ann. Inst. Super. Sanita.*, **22**, 1099–100.
Gilliland, S. E. (1979). *J. Food Protect.*, **42**, 164–7.
Gilliland, S. E. (1989). *J. Dairy Sci.*, **72**, 2483–94.
Gilliland, S. E. & Kim, H. S. (1984). *J. Dairy Sci.*, **67**, 1–6.
Gilliland, S. E. & Lara, R. C. (1987). *Appl. and Environ. Microbiol.*, **54**, 898–902.
Gilliland, S. E. & Speck, M. L. (1977). *J. Food Protect.*, **40**, 820–3.
Gilliland, S. E., Nelson, C. R. & Maxwell, C. (1985). *Appl. Environ. Microbiol.*, **49**, 377–81.
Goldin, B. R. & Gorbach, S. L. (1984). *Am. J. Clin. Nutr.*, **39**, 756–61.
Hood, S. K. & Zottola, E. A. (1987). *J. Food Sci.*, **52**, 791–2, 805.
Hull, R. R., Roberts, A. V. & Mayes, J. J. (1984). *Aust. J. Dairy Tech.*, **39**, 164–6.
Hunger, W. (1985). *Deutsche Molkerei-Zeitung*, **106**, 826, 828, 830, 832–3.
Hylmar, B. & Pokorna, L. (1986). *Prumysl Potravin*, **37**, 304–8.
Johnson, M. C., Ray, B. & Bhowmik, T. (1987). *J. Ind. Microbiol.*, **2**, 1–7.
Kanda, H., Wang, H. L. & Hesseltine, C. W. (1978). US Patent No. 4 066 792.
Keating, K. (1985). *Cult. Dairy Prod. J.*, **20**(2), 13–17.
Kim, H. S. (1984). *Korean J. Vet. Res.*, **24**, 149–62.
Kim, H. S. (1988). *Cult. Dairy Prod. J.*, **23**(3), 6–9.
Klaenhammer, T. R. (1985). *J. Dairy Sci.*, **65**, 1339–49.
Kleeman, E. G. & Klaenhammer, T. R. (1982). *J. Dairy Sci.*, **65**, 2063–9.
Kolars, J. S., Levitt, M. D., Aouji, M. & Savaiano, D. A. (1984). *New Engl. J. Med.*, **310**, 1–3.
Lidbeck, A., Gustafsson, J. A. & Nord, C. E. (1987). *Scand. J. Infect. Dis.*, **19**, 531–7.
Litschgi, M. S., Da Rugna, D., Mladenovic, D. & Grcic, R. (1980). *Fortschr. Med.*, **98**, 1624–7.
Luckey, T. D., Onderdonk, A. B., Finkelstein, R. A., Savage, D. C. & Wyatt, R. G. (1983). Intestinal Microecology. Prog. Food Nutr. Sci. 7, 7th International Symposium on Intestinal Microecology.
Maeyrae-Maekinen, A., Manninen, M. & Gyllenberg, H. (1953). *J. Appl. Bacteriol.*, **55**, 723–37.

Mantere-Alhonen, S. & Makinen, E. (1987). *Meijeritieteellinen Aikakauskirja*, **45**, 49–61.
Mautone, A., Monno, R., Palmieri, P. & Brandonisio, O. (1982). *Minerva Pediatr.*, **34**, 245–50.
McDonough, F. E., Wong, N. P., Wells, P., Hitchings, A. D. & Bodwell, C. E. (1985). *Nutr. Rep. Int.*, **31**, 1237–45.
Meszaros, L. A. (1989). *Chemtech*, January.
Miyamoto, T., Hirata, N. & Nakae, T. (1989). *Japanese J. Zootechnical Sci.*, **58**, 754–63.
Nabukhotny, T. K., Cherevko, S. A., Samigullina, F. I. & Grushko, A. I. (1983). *Voprosy Pitaniya*, **6**, 27–30.
Nahaisi, M. H. & Robinson, R. K. (1985). *Dairy Industries International*, **50**, 16–17.
Perdigon, G., Alvarez, S., Nader de Macias, M. E., Margni, R. A., Oliver, G. & Pesce de Ruiz Holgado, A. A. *Milchwiss.*, **41**, 344–8.
Pettersson, L., Graf, W. & Sewelin, U. (1983). *Symp. Swed. Nutr. Found.*, **15**, 127–30.
Pollman, D. S., Kennedy, G. A., Koch, B. A. & Allee, G. L. (1984). *Nutr. Rep. Int.*, **29**, 977–82.
Porubcan, R. S. & Sellars, R. L. (1979). *Microbial Technology*, 2nd Edn, Vol. 1, ed. H. J. Peppler & D. Pearlman. Academic Press, New York, p. 59.
Prajapati, J. B., Shah, R. K. & Dave, J. M. (1986). *Cultured Dairy Products J.*, **21**, 16–17, 20–21.
Prajapati, J. B., Shah, R. K. & Dave, J. M. (1987). *Aust. J. Dairy Technol.*, **42**, 17–21.
Reddy, G. V., Shahani, K. M., Friend, B. A. & Chandan, R. C. (1983). *Cultured Dairy Products J.*, **18**, 15–19.
Reddy, N. R., Roth, S. M., Eigel, W. N. & Pierson, M. D. (1988). *J. Food Protect.*, **51**, 66–75.
Robinson, R. K. (1987). *Suid Afrikaanse Tydskrif Vir Suiwelkunde*, **19**, 25–7.
Sandine, W. E., Muralidhara, K. S., Elliker, P. R. & England, D. C. (1972). *J. Milk and Food Tech.*, **35**, 691–702.
Savage, D. C. (1978). *Am. J. Clin. Nutr.*, **31**, 131–5.
Schaffner, D. W. & Beuchat, L. R. (1986). *Appl. Environ. Microbiol.*, **51**, 1072–6.
Sellars, R. L. (1967). *Microbial Technology*, ed. H. J. Peppler. Van Nostrand-Reinhold, Princeton, New Jersey, pp. 34–75.
Sharma, D. K. & Prasad, D. N. (1986). *Cultured Dairy Products J.*, **21**, 13–14.
Silles, A. & Hilton, T. (1987). *J. Food Protect.*, **150**, 812–14.
Solomadina, L. V. (1984). *Molochnaya Promyshlennost*, **9**, 42–4.
Speck, M. L. (1978). *J. Food Protect.*, **41**, 135–7.
Subramanian, P. & Shankar, P. A. (1985). *Cultured Dairy Products J.*, **20**, 17, 20–1, 24, 26.
Sudoma, A. L. US Patent Pending.
Watkins, B. A. & Miller, B. F. (1983). *Poult. Sci.*, **62**, 2152–7.
Wooley, R. E., Gilbert, J. P., Whitehead, W. K., Shotts, Jr. E. B. & Dobbins, C. N. (1981). *Am. J. Vet. Res.*, **42**, 87–90.

Chapter 6

THE HEALTH POTENTIAL OF PRODUCTS CONTAINING BIFIDOBACTERIA

J. A. KURMANN
Agricultural Institute Grangeneuve, Posieux, Switzerland

&

J. LJ. RAŠIĆ
Food Research Institute, Novi Sad, Yugoslavia

Bifidobacteria were isolated and described in the period 1899–1900 by Tissier, who named the type species *Bacillus bifidus communis* or *Bacillus bifidus*. These organisms were the predominant bacteria in the stools of breast-fed infants (Tissier, 1900). Since then, investigations have been concerned with various aspects of bifidobacteria, including their habitat and growth-promoting factors, nutritional significance and with devising culture media for the isolation and maintenance of strains. The opinion of Tissier that bifidobacteria were confined to breast-fed infants was accepted for a long time, but subsequently these organisms have been found in the stools of formula-fed infants and of adults.

In the past three decades, however, important progress has been made in the study of bifidobacteria in relation to their distribution, biology, taxonomy and significance in nutrition and health.

The potentially beneficial roles of bifidobacteria in the human intestine have been mentioned in many publications (Levesque *et al.*, 1959; Mayer, 1966; Schneegans *et al.*, 1966; Müting *et al.*, 1968; Poupard *et al.*, 1973; Bullen *et al.*, 1976; Mizutani & Mitsuoka, 1980; Rašić & Kurmann, 1983; Takano *et al.*, 1985). As a result, a variety of fermented milk products have been developed. These products include: fermented milks, fermented-milk beverages, buttermilk, sour cream, fresh cheese and baby foods.
This review will concentrate on the therapeutic aspects of fermented milks containing viable bifidobacteria.

HUMAN SPECIES OF THE GENUS

General

Bifidobacteria are normal inhabitants of the human intestine, but they are also found in the human mouth and vagina, and in the alimentary tract of various animals and honey bees; they also occur in sewage. They have characteristic morphology, physiology, biochemical characters, cell-wall constituents and DNA-base composition. The diagnostic characteristics of the genus are shown in Table I.

There are human and non-human species and biovars. The details of their biology are given elsewhere (Rašić & Kurmann, 1983), and details of their taxonomy are given by Scardovi in Bergey's Manual (Scardovi, 1986).

TABLE I
DIFFERENTIATING CHARACTERISTICS OF THE GENUS *BIFIDOBACTERIUM*

Morphology:	Non-spore-forming and non-motile rods; variable in appearance; Gram-positive; methylene blue may stain internal granules, but not the entire cell.
Physiology:	Anaerobic.
Biochemical characters:	Major products of glucose fermentation; acetic and L(+)-lactic acids in a molar ratio of 3:2; additional acetic and formic acids usually produced by the splitting of pyruvate, not fast-acid producers; gas (CO_2) not produced from glucose, negative tests for: catalase, nitrate reduction (except when bacteria grown in the presence of lysed red cells), formation of indole and liquefaction of gelatin; acid not produced from: rhamnose, sorbose, adonitol, dulcitol, erythritol and glycerol; positive for fructose-6-phosphate phosphoketolase in cellular extracts.
Cell wall composition:	The cell-wall peptidoglycan is not uniform. The basic amino acid in the tetrapeptide can be either ornithine or lysine; various types of cross-linkage (single amino acid residue or peptide).
Phospholipid composition:	Specific phospholipids: polyglycerolphospholipid and its lyso derivatives, alanyl phosphatidylglycerol, and lyso derivatives of diphosphatidylglycerol.
DNA-base composition:	Guanine plus cytosine (G+C) content: 55–67 M%.

Source: Rašić & Kurmann (1983) and Scardovi (1986).

HEALTH POTENTIAL OF PRODUCTS CONTAINING BIFIDOBACTERIA 119

Relevant Species

In the past, bifidobacteria were regarded as one species *Lactobacillus bifidus*, but subsequently they were designated as a separate genus *Bifidobacterium*, comprising eleven species (Buchanan & Gibbons, 1974). Recently, *Bifidobacterium* has been classified into twenty-four species: nine occurring in humans, twelve in animals and three in honey bees (Scardovi, 1986).

The human species include: *Bif. bifidum* biovars *a* and *b*, *Bif. longum* biovars *a* and *b*, *Bif. infantis*, *Bif. breve*, *Bif. adolescentis* biovars *a*, *b*, *c* and *d*, *Bif. angulatum*, *Bif. catenulatum*, *Bif. pseudocatenulatum* and *Bif. dentium*.

The optimum growth temperature is 37–41°C, maximum 43–45°C and minimum 25–28°C. The optimum pH for initial growth is 6·5–7·0, and at 4·5–5·0 or 8·0–8·5, no growth occurs. All human species are able to utilize glucose, galactose, lactose and, generally, fructose as the carbon sources, and ammonia as the sole source of nitrogen. Many strains require riboflavin (except those of *Bif. bifidum*) and pantothenate or pantethine for growth, whereas requirements for other vitamins vary among different species.

In the manufacture of fermented milks, *Bif. bifidum* is the species most often used, followed by *Bif. longum* and *Bif. breve*. They are usually used in combination with other lactic acid bacteria due to their slow acid production. Some pharmaceutical preparations contain *Bif. infantis* as well, often in combination with other lactic acid bacteria. Differentiating characteristics of these species are shown in Table II.

As shown in Table II, *Bif. bifidum* has a limited fermentation ability; biovar *a* (from adults) slowly ferments sucrose and melibiose, whereas biovar *b* (from infants) does not ferment sucrose but ferments melibiose. In contrast to other species, *Bif. bifidum* does not ferment raffinose, and most strains are maltose-negative. It is weakly ureolytic. *Bif. longum* is distinguished by the fermentation of pentoses, but not cellobiose. Biovar *a* (from adults) slowly ferments mannose, and biovar *b* (from infants) is mannose-negative. It is the only human species having a variety of plasmids, but no phenotypic properties have been correlated with them (Sgorbati *et al.*, 1982). More than 90% of strains are urease-negative. *Bif. breve* and *Bif. parvulorum* have homologous DNA and are merged under *Bif. breve*. It ferments ribose, but not arabinose and xylose. More than 90% of strains are urease-negative; found in the stools of infants and adults. *Bif. infantis*, *Bif. lactentis* and *Bif. liberorum* are genetically homologous and are merged under *Bif. infantis* (synonym *Bif. para-*

TABLE II
DIFFERENTIATING CHARACTERISTICS OF *BIF. BIFIDUM*, *BIF. LONGUM*, *BIF. BREVE* AND *BIF. INFANTIS*

Characteristics	Bif. bifidum	Bif. longum	Bif. breve	Bif. infantis
Morphology in TPY agar stabs[a]	amphora-like cells	elongated and thin cells	short and thin rods	no specific traits
Acid from				
D-Ribose	−	+	+	+
L-Arabinose	−	+	−	−
Lactose	+	+	+	+
Cellobiose	−	−	d	−
Melezitose	−	+	d	−
Raffinose	−	+	+	+
Sorbitol	−	−	d	−
Xylose	−	d	−	d
Mannose	−	d	+	d
Fructose	+[b]	+	+	+
Sucrose	c,d	+	+	+
Maltose	−[d]	+	+	+
Trehalose	−	−	d	−
Melibiose	d	+	+	+
Mannitol	−	−	d	−
Inulin	−	−	d	d
Salicin	−	−	+	−

[a]TPY agar (Scardovi, 1986).
[b]Few strains do not ferment this sugar.
[c]When positive it is fermented slowly.
[d]Some strains ferment this sugar.
Source: Scardovi (1986).

bifidum), but they differ in their ability to ferment xylose and other sugars. This species is genetically related to *Bif. longum*. It is urease-positive, and is found in the stools of infants.

These four relevant species, as regards their use in fermented milks, are not 'fast-acid' organisms (a characteristic of bifidobacteria), but with repeated transfers in culture media, their multiplication rate and acid production may be increased. Acid production is increased particularly following the addition of growth-promoting substances, e.g. yeast extract or autolysate, cysteine, pepsin or pepsin-digested milk, whey protein, etc. A freshly prepared, milk culture of *Bif. bifidum*, for instance, is acidified and often coagulated in 18–96 h at 37–38°C. The acid producing ability of bifidobacteria is also dependent on strain characteristics.

Some strains of bifidobacteria show weak proteolytic activity in milk, but there are considerable differences in the activity of strains (Rašić & Kurmann, 1983).

Bifidobacteria may survive, to a certain extent, the acidity of gastric juice and pass into the small intestine. Various strains differ in their acid tolerance, and the ingestion of large numbers of bifidobacteria may consequently increase their survival (Schuler-Malyoth et al., 1968; Lipinska, 1978). They are sensitive to increased levels of bile acids, e.g. deoxycholic acid has been shown to be bacteriostatic to bifidobacteria at levels of 0·02–0·05% and bactericidal at 0·2–0·5%. The concentration of deoxycholate in the normal faeces varies from 0·05 to 0·2% (Catteau et al., 1971). The level of faecal bile acids is influenced by the amount of fat in the diet. Fatty diets may increase the amounts of bile acids in the faeces, and consequently increase the inhibitory effect on bifidobacteria.

Occurrence and Species Distribution

Bifidobacteria are the predominant organisms in the large intestine of breast-fed infants accounting for about 99% of the faecal microflora (10^9–10^{11} colony forming units/g faeces), whereas enterobacteria, enterococci and lactobacilli comprise about 1% (range 1–15%) of the flora, and bacteroides, clostridia and other micro-organisms are absent or insignificant (Frisell, 1951; Hoffmann, 1966; Gorbach et al., 1967; Mata et al., 1969; Haenel, 1970; Stark & Lee, 1982; Roberts et al., 1985). The pH values of the stools range from 5·0 to 5·5 (see Fig. 1).

The faecal microflora of formula-fed infants resembles that of adolescents and adults. Bifidobacteria are approximately equal in numbers to, or may be outnumbered by, bacteroides including other species, and most authors agree that the numbers of micro-organisms (other than bifidobacteria) are larger in the stools of formula-fed infants than in breast-fed infants (Frisell, 1951; Haenel et al., 1970; Stark & Lee, 1982; Roberts et al., 1985; Benno & Mitsuoka, 1986). The pH values of the stools range from 6·4 to 7·0 or higher.

In adolescents and adults, bifidobacteria are a major component of the large intestinal microflora, but they are usually outnumbered by bacteroides, including eubacteria, peptostreptococci and ruminococci, whereas lactobacilli, enterococci and enterobacteria are a smaller component of the flora. Bifidobacteria are reduced significantly in the stools of old people, and clostridia, enterobacteria and enterococci are increased. This is usually due to a diminished secretion of gastric juice in this age group (Orla-Jensen *et al.*, 1945; Hoffmann, 1966; Benno & Mitsuoka, 1986).

These micro-organisms are also present in the small intestine, but in much smaller numbers than in the large intestine and faeces.

The population of bifidobacteria in the intestine of breast-fed infants is relatively unstable, and small influences, e.g. changes in nutrition, common infections, vaccinations, etc., may upset the microbial balance. In adults, the microflora in the lower intestine is more complex and stable, but drastic influences may disturb the normal microbial balance.

The distribution of species and biovars of bifidobacteria in the intestine is influenced by the host physiology, age and diet. *Bif. bifidum* biovar *b*, *Bif. longum* biovar *b*, *Bif. infantis* and *Bif. breve* occur in the stools of

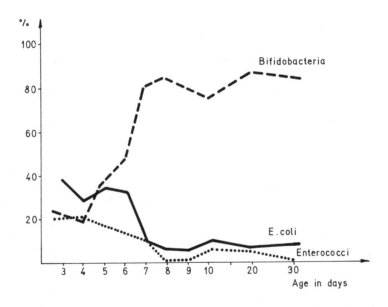

FIG. 1. The proportions (in per cent) of bifidobacteria, *E. coli,* and enterococci in the faecal flora of breast-fed infants within 30 days of birth. Clostridia were often found in the faeces within the first 5 days, and thereafter nearly completely disappeared. Bacteroides were found in each 5th sample only (Hoffmann, 1966).

infants. There is no general agreement about the species which dominates in the stools of breast-fed and formula-fed infants. Earlier reports showed that *Bif. infantis* is the predominating species in the stools of breast-fed infants (Dehnert, 1957; Reuter, 1963; Petuely & Lindner, 1965; Werner, 1966; Mitsuoka, 1969; Haenel, 1970). Subsequently, it has been shown that *Bif. bifidum* is the most commonly isolated species from the stools of breast-fed infants, whereas *Bif. longum* is the most common in the stools of formula-fed infants (Beerens *et al.*, 1980; Grütte & Müller-Beuthow, 1980). Other reports have indicated that *Bif. breve* is the most common species in the stools of both breast-fed and formula-fed infants, whereas *Bif. bifidum* was found to be more common in the stools of breast-fed infants than those of formula-fed infants (Yuhara *et al.*, 1983; Mitsuoka, 1984; Benno & Mitsuoka, 1986; Mutai & Tanaka, 1987).

Bif. longum biovar *a* and *Bif. adolescentis* are typical of adults. However, *Bif. bifidum* biovar *a*, *Bif. catenulatum* and *Bif. pseudocatenulatum* may also occur (Biavati *et al.*, 1986). In the stools of old people, *Bif. adolescentis* biovar *b* usually occurs in larger numbers than in other age groups, but species such as *Bif. bifidum* and *Bif. longum* may also be found in small numbers (Mitsuoka, 1984; Mutai & Tanaka, 1987).

HEALTH PROPERTIES ASSOCIATED WITH THE PRESENCE OF BIFIDOBACTERIA IN THE INTESTINE

Bifidobacteria in the Intestine of Infants and Small Children

Table III reviews the potentially beneficial roles of bifidobacteria in the intestine of infants and small children.

TABLE III
THE POTENTIAL ROLES OF BIFIDOBACTERIA IN THE INTESTINE OF INFANTS AND SMALL CHILDREN

Indigenous organisms	Dietary adjuncts
Competitive antagonism against invading pathogens	Modification of the intestinal flora in formula-fed infants
Antimicrobial substances produced —organic acids —other substances	Prophylactic effects Supportive therapy of intestinal infections and disturbances
Improved N retention and weight gain in infants	Beneficial competition with other intestinal bacteria
Inhibition of nitrate reduction	

Functions of Indigenous Bifidobacteria
The beneficial functions of bifidobacteria in the intestine result from a number of their biological activities. The best known are described below.

Competitive Antagonism against Invading Pathogens
Large numbers of bifidobacteria may prevent colonization of the intestine by invading pathogens, both through competition for nutrients and for attachment sites to the epithelial surfaces. Normally, the indigenous intestinal microflora, including bifidobacteria, acts synergistically with its host's immunological system in protecting against infections by intestinal pathogens.

Production of Organic Acids
Bifidobacteria produce acetic and lactic acids and a small amount of formic acid—which lower the pH of the large intestine and, thereby,

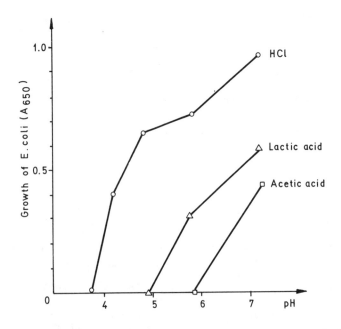

FIG. 2. Growth inhibiting activity of acids to *E. coli*. Acid: 0·1 M, and pH was adjusted with NaOH; culture: 37°C for 30 h (Tamura, 1983).

inhibit the growth of undesirable micro-organisms. Acetic acid has a stronger antagonistic effect against Gram-negative bacteria than lactic acid, and the former is produced in greater amounts by bifidobacteria. It has been shown, for example, that the minimum pH values at which salmonellae commence growth (under laboratory conditions) are 5·40 for acetic acid, 4·40 for lactic acid and 4·05 for citric and hydrochloric acids (Chung & Goepfert, 1970).

Similar results have been reported for the growth-inhibiting activity of acetic, lactic and hydrochloric acids against *Escherichia coli* (Tamura, 1983) (Fig. 2). Organic acids present in their free (undissociated) forms may also have a direct, destructive effect on other micro-organisms (Poupard *et al.*, 1973). Diet considerably influences the level of acids produced, as well as their forms. Feeding infants with human milk favours a predominance of bifidobacteria, an absence of putrefactive micro-organisms and acidic stools (pH 5·0–5·5). The presence of a buffer consisting of acetic acid and acetate was shown in the faeces of breast-fed infants, but not in those from formula-fed infants (Bullen & Tearle, 1976; Bullen *et al.*, 1976).

Feeding with breast milk also supplies the newborn with protective factors, such as specific antibodies, macrophages and lymphocytes, lysozyme, lactoferrin, lactoperoxidase and volatile fatty acids (Braun, 1971; Bullen *et al.*, 1972; Hambraeus, 1978; Klaus & Dias-Rossello, 1980; Jelliffe & Jelliffe, 1981). Many such protective factors are present only in traces in cows' milk.

Feeding infants with formula feeds gives rise to a large intestinal microflora resembling that of adults, and the stools have a neutral to alkaline pH. Improved formulations of milk (high lactose, low protein, low calcium phosphate contents, together with a low buffering capacity) and the incorporation of appropriate bifidogenic substances in the diet induce the growth of bifidobacteria, and also decrease the pH in the large intestine.

Antimicrobial Substances other than Organic Acids

Although organic acids play a major role in the antagonistic effect of bifidobacteria against many micro-organisms, some reports indicate that bifidobacteria may produce specific antibacterial substances. The possibility of lysozyme production by bifidobacteria has been suggested on the grounds that egg-white lysozyme and a metabolite of bifidobacteria may have common antigenic determinants (Minigawa, 1970), but more evidence is needed to confirm this observation. Recently, antibacterial activity of some strains of *Bif. bifidum* has been reported. Maximum inhibition was against *Micrococcus flavus*, followed by *Staphylococcus aureus*, *Bacillus cereus*, *Escherichia coli*, *Pseudomonas fluorescens*, *Salmonella typhosa* and *Shigella dysenteriae*.) The greatest inhibitory effect was in the pH range 4·8–5·5, but above pH 5.5 it was lost. The partially-purified, inhibitory substance named 'Bifidin' was found to contain amino acids, principally phenylalanine and glutamic acid (Anand *et al.*, 1984; 1985). The antibacterial effect of some strains of *Bif. bifidum* may be effective mainly in breast-fed infants whose stools have a pH of 5·0–5·5.)

The protective effect of bifidobacteria and their high molecular weight compounds on epithelial cells against invasion and/or multiplication of *Shigella* spp. was shown using a tissue culture infection assay. *Bif. infantis* and its high molecular weight substances had the greatest protective effect against *Shigella flexneri* 5503–01, followed by *Bif. breve*, *E. coli* and *Streptococcus faecalis* (Nakaya, 1984; Okamura *et al.*, 1986). It may be supposed that this is the same *in vivo*.

Improved N Retention and Weight Gain in Infants
Bifidogenic substances, such as lactulose (Stenger & Wolf, 1962; Horečny, 1964) and purified pig's gastric mucin (Inoue & Nagayama, 1970), added to the diet of formula-fed infants have been reported to increase N retention and weight gain, and to reduce faecal pH.

Inhibition of Nitrate Reduction
The ability of bifidobacteria to inhibit the reduction of nitrate may be important in feeding infants. The reduction of nitrate to nitrite is favoured by the gastric juice with a pH greater than 4·0, and by the presence of nitrate-reducing bacteria in the intestine (Fomon, 1974). Nitrite ions oxidize the ferrous-ions of haemoglobin to ferric-ions, and the resulting compound, methaemoglobin, is incapable of binding molecular oxygen. Bifidobacteria inhibit the formation of nitrites mainly through an inhibitory effect on the growth of undesirable bacteria. Consequently, the use of bifidobacteria in baby foods may reduce the incidence in infants of alimentary methaemoglobinaemia caused by nitrites (Doležalek, 1979; Brežina *et al.*, 1983).)

Dietary Adjuncts
The potentially beneficial functions of bifidobacteria in the intestine of infants and small children have led to their suggested use as dietary

adjuncts. *Bif. bifidum* is the species most often used, but, in some cases, strains of *Bif. longum* and *Bif. infantis* are used as well. Bifidobacteria may be used as such, or together with their appropriate growth-promoting substances, for the following purposes: (a) to modify the intestinal microflora of formula-fed infants; (b) to protect against intestinal infections or side-effects of antibiotic therapy; (c) to support therapy of enteric infections and intestinal disturbances (Rašić, 1983).

Modification of the Intestinal Microflora
It may be possible to modify the intestinal flora of formula-fed infants if: (a) large numbers of viable cells of *Bif. bifidum* biovar *b* (10^8–10^9 cells/day) are fed; (b) they survive gastric transit and, preferably, can adhere to epithelial intestinal surfaces and grow; and (c) a fermentable carbohydrate is available to the cells in the large intestine.

Studies of infants fed formulas containing a freeze-dried culture of *Bif. bifidum* (10^7 cells/day) have shown the effective implantation in most cases, but at levels below those found in breast-fed infants (Levesque *et al.*, 1959; Schneegans *et al.*, 1966). Since the lactose in formula feeds does not reach the large intestine in sufficient amounts to promote the growth of bifidobacteria, the addition of bifidogenic substances to the diet could provide a fermentable carbohydrate; improved formulation of the milk is also an important factor. Since the extracellular polysaccharide fraction of bifidobacteria has been shown to participate in the adherence of bacterial cells to epithelial surfaces (Sato *et al.*, 1982), the use of polysaccharide-producing strains in baby foods may be preferred.

Prophylactic Effects
A protective effect of bifidobacteria against enteric infections has been demonstrated in human studies. Comparative feeding of two groups of infants—one group (114 infants) receiving a formula feed containing a culture of *Bif. bifidum* and lactulose, and the other group (78 infants) receiving a buttermilk preparation—showed that enteric infections occurred once in the first group and eight times in the second one (Kaloud & Stögmann, 1968).

The antagonistic relationships between bifidobacteria and enterobacteria in the digestive tract have been shown in the following animal studies. The administration of a 'bifidus' milk containing 10^7 cells/ml of *Bif. longum* to mice at a level of 0·2 ml per day, for 14 days, induced a suppression of enterobacteria, and an increase in the population of bifidobacteria (Momose *et al.*, 1982). The acidic intestinal environment created by bifidobacteria may account, in part, for this effect. The feeding of 28-day-old albino rats with a spray-dried, infant formula containing 1.55×10^6 colony forming units/g of *Bif. bifidum* for 12 days, increased faecal counts of *Bif. bifidum* to 1.2×10^6 colony forming units/g on the eighth day of feeding, whereas counts of *E. coli* decreased to nil in the same period (Pahwa & Mathur, 1982). The antagonistic interaction between *Bif. bifidum* and *Proteus vulgaris* was shown in studies on gnotobiotic chicks (Timoshko *et al.*, 1979). The use of a dried concentrate of *Bif. bifidum* in the control of gastro-intestinal disease in 2 to 3-day old piglets and broiler chicks was shown to reduce significantly the incidence of disease and deaths (Ervolder *et al.*, 1984). Similar results have been obtained in lambs weaned at 2–3 days, and fed with a bifid preparation containing *B. bifidum* and *Lac. acidophilus* (Solomonov *et al.*, 1984).

The results of studies obtained with animals may not extrapolate directly to human beings. However, they may have some indicative importance. The significance of using 'host specific' micro-organisms for the preparation of fermented milks for dietetic use has been particularly pointed out by Gurr *et al.* (1984).

Bifidobacteria may be used to prevent side-effects of antibiotic therapy. It has been shown, that the oral administration of a bifidus milk preparation containing viable *Bif. bifidum* and its growth factors to an infant, following the discontinuation of penicillin therapy, increased the population of bifidobacteria in the faeces and suppressed the growth of *Candida albicans* (Mayer, 1966). Another report refers to the administration of a preparation containing antibiotic-resistant *Bif. bifidum* and *Lac. acidophilus* to children (between 13 days and 8 years of age) during long-term, broad-spectrum antibiotic therapy for a variety of conditions. The treatment prevented the occurrence of dyspeptic disturbances, and the administered organisms were detected in the faeces of 50% of the patients, even several days after termination of their administration (Branca *et al.*, 1979).

The Potentially Therapeutic Effects

Some evidence suggests that dietary bifidobacteria produce beneficial effects in infants and children with enteric infections and gastro-intestinal disturbances. It has been shown that the oral administration of a culture of *Bif. bifidum* in conjunction with a dietetic regimen produced beneficial effects in infants with bacterial enterocolitis. The feed consisted of a high-fluid diet, followed by cream of rice and human milk (Tasovac, 1964). Likewise the oral administration of a freeze-dried culture of *Bif.*

bifidum in conjunction with lactulose has been shown to eradicate enteropathogenic *E. coli* strains in more than 80% of cases (Schneegans *et al.*, 1966).

The beneficial effect of bifidobacteria, when used in conjunction with lactulose (or comparable dietetic feeding), against enteric infections indicates the significance of a low pH in the large intestine, and the influence of the feed.

A recent report indicates the therapeutic roles of dietary bifidobacteria in dyspeptic conditions, such as intractable diarrhoea attributable to antibiotic therapy (Mutai & Tanaka, 1987). Fifteen patients, between 1 month and 15 years of age (a mean of 2·5 years), received antibiotics, such as cephems, penicillins and aminoglycosides for the treatment of septicaemia and respiratory tract infections. During treatment, diarrhoea appeared and lasted for 1–10 weeks. The numbers of bifidobacteria decreased in the faeces, and those of *Candida* spp. and *Enterococcus* considerably increased. The oral administration of a *Bifidobacterium* preparation (bifidus yoghurt) improved the stool frequency within 3 to 7 days in all patients, and this was accompanied with the restoration of the intestinal microflora. The nutritional benefits of bifidus yoghurt have been recently reviewed in more detail (Rašić, 1987a).

Treatment of gastro-intestinal diseases, such as chronic gastroduodenitis and peptic ulcer, in 21 children with an 'Antacid bifilakt' preparation containing bifidobacteria and lactobacilli (species not specified) and lysozyme has been reported to be highly effective (Dorofeichuk *et al.*, 1983).

Beneficial Competition with Other Intestinal Bacteria

A large proportion of bifidobacteria in fermented milks may survive the gastric transit (Schuler-Malyoth *et al.*, 1968; Lipinska, 1978; Gilliland, 1979), and may thus compete beneficially with other intestinal bacteria; consequently, they may be successful in aiding indigenous bifidobacteria and lactobacilli to combat invading pathogens.

Bifidobacteria in the Intestine of Adults

Influence on the Intestinal Microflora

It is difficult to induce major changes in the indigenous intestinal microflora by alterations in the diet, because the composition of the flora in the lower intestine is stable in healthy adults. However, the regular consumption of fermented milks containing viable bifidobacteria and lactobacilli ensures a continued supply of starter bacteria to the intestine, and these organisms help to maintain a proper balance of the resident flora or to correct a disturbed balance.

The metabolic activity of the intestinal microflora may be affected by alterations in the diet, and the regular consumption of products containing bifidobacteria and lactobacilli may mitigate certain harmful effects of some components of the intestinal microflora.

It has been shown that feeding with a diet containing large numbers of viable bifidobacteria induces changes in some components of the intestinal flora, and in their metabolic activities, during the period of feeding. The administration of *Bif. breve* (10^{10} cells/day) to rats increased the numbers of bifidobacteria in the faeces and decreased enterobacteria and bacteroides, together with reductions of urinary indican, cadaverine, piperidine and indole acetate, also ammonia in the portal blood. Similar results have been reported in humans given *Bif. breve* at 10^9 cells/day (Mada, 1981).

Likewise the administration of *Bif. breve* at 10^9 cells/day, together with a growth-promoting factor of a transgalactosylated oligosaccharide (TOS) at 0·75 g/day, to rats with abnormal metabolic activities of the intestinal flora has been reported to give effective results. The abnormal metabolic activities of the flora were induced by feeding a diet supplemented with 3% lysine–1% tryptophan or 40% egg-white protein to rats with human or conventional floras, respectively. As a result of the introduction of bifidobacteria, the concentrations of indican, cadaverine, piperidine and indole-3-acetate in the urine, and ammonia in the portal blood decreased, along with increased numbers of bifidobacteria in the faeces and decreased numbers of enterobacteria and clostridia. The host hepatic function, such as aniline hydroxylase and aminopyrine-*N*-demethylase activities, was improved as well (Tohyama *et al.*, 1982). However, it is unknown whether these responses would occur in human beings.

Some intestinal bacteria are able to nitrosate amines using nitrite as the nitrosating agent. This reaction leads to the formation of nitrosamines, which have been shown to be potent carcinogens. However, the synthesis of nitrosamines, under conditions in which bacterial activities are possible, may be reduced by enzymatic degradation carried out by some strains of lactobacilli and bifidobacteria (Rowland & Grasso, 1975).

Prophylactic and Therapeutic Aspects

Drastic influences, such as gastric achlorhydria or hypochlorhydria, disorders of intestinal motility, stagnation of intestinal contents, disorders of immunological systems and some diseases, induce changes in the gastrointestinal microflora. They are characterized by microbial 'overgrowth' of the stomach and small intestine, and by a disturbed, large intestinal flora.

Certain antibiotics, irradiation of the abdomen with gamma or X-rays, and stress conditions may also disturb the balance of the intestinal microflora, leading to a reduction or disappearance of lactobacilli and bifidobacteria.

The consumption of bifidobacteria and lactobacilli in fermented milks may help in the restoration of the intestinal microflora, particularly when combined with the simultaneous elimination of influences causing the disturbed balance. The potentially beneficial role of dairy foods containing bifidobacteria and acidophilus bacteria in nutrition and health has been reviewed elsewhere (Rašić, 1983; 1984; Rašić & Kurmann, 1983).

Good results have been reported in the treatment of intestinal disturbances after oral antibiotic therapy (Leperchey & Frottier, 1965; Bamberg, 1966) and irradiation therapy (Haller & Kräubig, 1960; Neumeister & Schmidt, 1963) using preparations containing *Bif. bifidum* or *Bif. longum* and *Lac. acidophilus*.

It has been shown that the oral administration of kanamycin (40 mg/kg) to adult volunteers for five days considerably decreased the populations of bifidobacteria, lactobacilli and enterococci, and increased enterobacteria. Three oral administrations of autostrains of bifidobacteria and lactobacilli, after the termination of the antibiotic intake, rapidly restored the intestinal microflora (Korshunov *et al.*, 1985).

A disturbed balance of the intestinal microflora in patients with leukaemia has been reported to be improved by the oral administration of bifidobacteria and acidophilus bacteria (Kageyama *et al.*, 1984). Drinking water was substituted by milk containing *Bifidobacterium* (species not specified) and *Lac. acidophilus*. Two hundred millilitres of milk containing about 2×10^9 of bifidobacteria and 2×10^9 of acidophilus bacteria were given to 28 patients every day for 3 months; and 28 patients were controls. The numbers of *Klebsiella* spp., *Pseudomonas* spp., *Proteus* spp., *Citrobacter* spp. and *Candida* spp. in the faeces of patients receiving bifidobacteria preparation decreased significantly compared with patients not receiving the bifidobacteria preparation (see Tables IV and V); also the concentrations of urine indican and blood endotoxin were comparatively reduced in patients receiving bifidobacteria preparations.

As mentioned earlier, the beneficial effects of ingested bifidobacteria may be obtained if: (a) large numbers of viable cells (10^8-10^9 cells/day)

TABLE IV
CHANGES IN MINOR MEMBERS OF INTESTINAL BACTERIA DURING ANTILEUKAEMIC THERAPY[a]

Organism	Control (10 cases)	Treatment with antileukaemic drug (56 cases)	
		Without Bifidobacteria (28 cases) (%)	With Bifidobacteria (28 cases) (%)
Klebsiella	0	3 (10·7)	2 (7·1)
Citrobacter	0	6 (21·4)	5 (17·9)
Pseudomonas Proteus vulgaris	0	8 (28·6)	2 (7·1)

[a]Cases in which more than 10^6 colonies/g faeces were observed.
Source: Kageyama et al. (1984).

are introduced; (b) they survive gastric transit and preferably adhere to epithelial surfaces and grow; and (c) a fermentable carbohydrate is available to the cells in the intestine. The antagonistic effect of bifidobacteria against undesirable micro-organisms may be due to the acetic and lactic acids they produce and, possibly, to their effect on the immunosystem of the host.

Many reports have indicated the beneficial roles of bifidobacteria in the management of chronic liver disease, especially when used in conjunction with their growth-promoting substances (Müting et al., 1968; Ordnung &

TABLE V
EFFECT OF *BIFIDOBACTERIUM* ADMINISTRATION ON NUMBERS OF *CANDIDA* IN FAECES OF PATIENTS RECEIVING ANTILEUKAEMIC THERAPY[a]

No. of Candida per g faeces	Administration of Bifidobacterium (16 cases)		Non-administration of Bifidobacterium (Control—11 cases)
	Before	After	
10^8	4	0	3
10^7	6	1	3
10^6	4	4	3
10^5	2	3	2
$< 10^4$	—	8	0

[a]Examination of 27 patients proved 10^5 or more *Candida*/g faeces.
Source: Kageyama et al. (1984).

Müting, 1973; Müting, 1977; Koizumi et al., 1980). The dietary administration of Bif. bifidum, together with lactulose, has been shown to assist in re-establishing the balance of the intestinal microflora which is usually disturbed in liver cirrhosis, and this is accompanied by decreasing the faecal pH and by a reduction of ammonia and free phenols in the blood (see Fig. 3) (Müting et al., 1968; Müting & Ordnung, 1976; Müting, 1977).

At a low colonic pH, much of the ammonia is in the form of the ammonium ion, which is not absorbed, as is its non-ionized form, by passive diffusion. The altered equilibrium of the non-ionized and ionized forms of ammonia may influence the movement of ammonia from the blood to the colon, thereby reducing the blood ammonia level. The ability of bifidobacteria to utilize ammonia as a source of nitrogen may also lead to a decrease of ammonia in the colon (Sahota et al., 1982; Scardovi, 1986).

The superior digestibility of bifidus milk, compared with the unfermented milk from which it is made, enables the patient to maintain his protein intake which is normally restricted. It is important to use non-urease producing strains of bifidobacteria for the manufacture of bifidus milk products.

The Potentially Antitumour Activity of Bifidobacteria

Studies with laboratory animals have indicated a possible antitumour effect of bifidobacteria. In experiments with BALB/c mice, Meth-A ascites tumour cells (25×10^3 cells) were transplanted, and a suspension of Bif. infantis (10^9 cells) was intraregionally injected into the mice one or two days after tumour inoculation, and repeated 4–6 times. No significant difference in tumour size between mice inoculated with tumour cells and the controls was observed on the eighth day, but later on, many established tumours completely or partially regressed in the experimental group of mice. The antitumour effect decreased when the first bacterial injection started three or seven days after tumour inoculation, or when mice were inoculated with a large number (500×10^3 cells) of tumour cells. The stimulation of the host immune response has been suggested to be involved in the antitumour effect of bifidobacteria (Kohwi et al., 1982).

Liver tumourigenesis has been reported to be reduced by the presence of Bif. longum in the gut of gnotobiotic C3H/He male mice (Mizutani & Mitsuoka, 1980). The effect may be due to the stimulation of immune response of the host, and/or to decreasing the activities of some faecal bacterial enzymes by bifidobacteria; more evidence is needed.

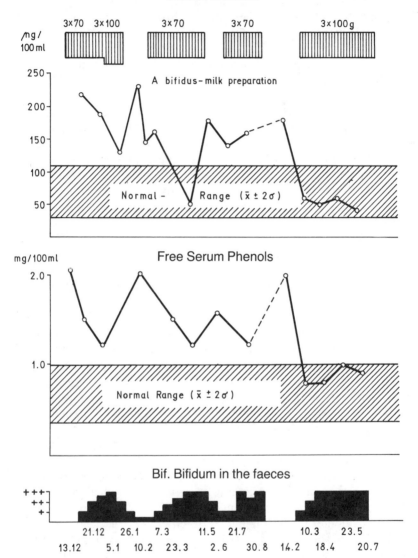

Fig. 3. The long-term treatment of decompensated liver cirrhosis with a bifidus milk preparation containing viable *Bif. bifidum* and lactulose (12-years-old patient) (From Ordnung & Müting, 1973).

Studies on rats have shown that chemically-induced colon tumourigenesis may be inhibited by feeding with sour milk. Male rats (F 344), six weeks of age, were fed with either a test diet: (a) sour milk (fermented with cultures of *Lac. helveticus* and *Candida utilis*); (b) artificially acidified milk; (c) starter cells; (d) a standard diet. From 6 to 22 weeks of age, nine rats in each group received, every week, an intraperitoneal injection of 1,2-dimethylhydrazine solution (20 mg/kg body weight), and were fed with the test diets for 10 weeks after the last injection of DMH. At 32 weeks of age, all rats were examined for the incidence of tumours. The smallest average number of colon tumours was found in rats fed with sour milk (see Table VI). However, the number of tumours in the small intestine did not differ significantly among rats fed with either diet (Takano *et al.*, 1985).

(The antitumour effect may be explained by the beneficial activities of bifidobacteria present in increasing numbers in the intestine of rats fed with sour milk.) Feeding studies on mice have shown that the life span of animals fed with a diet containing 14% of sour milk was, on average, 8% longer than those fed with either a diet containing 1·6% whole milk powder or a control diet; the numbers of bifidobacteria in the intestines of mice fed with sour milk were ten times higher than those fed with either whole milk or the control diet (Takano *et al.*, 1985).

Apparently, the potentially antitumour effect of bifidobacteria may involve various mechanisms, and further research is needed.

TABLE VI

INCIDENCE OF DMH-INDUCED COLON TUMOURS IN RATS FED A DIET CONTAINING SOUR MILK, ARTIFICIALLY ACIDIFIED MILK OR STARTER CELLS

Group	No. of rats	Animals with colon tumours	No. of colon tumours per rat		
			Total	Adenoma	Adenoma-carcinoma
Control	9	9	2·6	1·3	1·3
Sour milk	9	6	1·0[a]	0·4	0·6
Acidified milk	9	9	3·4	2·0	1·4
Starter cells	8	7	2·3	1·8	0·5

[a]Significantly different from control $p < 0.05$.
Source: Takano *et al.* (1985).

Effect on the Immune System of the Host

Current evidence suggests that allochthonous micro-organisms derived from food or other extraneous sources are more immunogenic to their hosts than are indigenous organisms of similar types in the gastro-intestinal tract (Savage, 1977; Moreau et al., 1978). Starter bacteria provide a significant source of antigens, which induce some immunological activity in their hosts. Experiments with mice have shown that supplementary feeding of yoghurt stimulated the enlargement of lymph nodes in the spleen. The numbers of B lymphocytes (associated with humoral immunity) and T lymphocytes (associated with cell-mediated immunity) were increased significantly, and also the serum immunoglobulin G (IgG_2); similar results have been obtained with germ-free mice (Conge et al., 1980; Lemonnier, 1984).

Experiments on laboratory animals have demonstrated the immunogenical effect of bifidobacteria as well. Germ-free mice monoassociated with *Bif. longum* (GB) were shown to live longer than non-treated, germ-free mice (GF) after intra-gastric or intra-venous administration of a lethal dose of *E. coli*, or intra-venous injection of endotoxin. The anti-lethal effect was induced only by viable *Bif. longum*, and occurred three weeks, but not less than 2 weeks, after association with *Bif. longum* (see Table VII). Since this effect was not observed in germ-free athymic mice,

TABLE VII

PROTECTIVE EFFECT OF *BIF. LONGUM*—MONOASSOCIATION AGAINST LETHAL INTRA-VENOUS CHALLENGE WITH *E. COLI*

Mice	Weeks after monoassociation	Dose of E. coli (viable units/mouse)	Mortality^a (%)
GB^b	2	5×10^8	5/5 (100·0)
		4×10^8	5/5 (100·0)
GF^b	2	5×10^8	5/5 (100·0)
		4×10^8	5/6 (83·3)
GB	3	5×10^8	2/5 (40·0)
		4×10^8	2/7 (28·6)^c
GF	3	5×10^8	5/6 (83·3)
		4×10^8	5/6 (83·3)

^{a}Dead/tested at 18 h after injection.
^{b}GB: germ-free mice monoassociated with *Bif. longum*; GF: germ-free mice non-treated.
^{c}$p < 0.01$ versus GF group.
Source: Yamazaki et al. (1982).

it has been suggested that bifidobacteria induce cell-mediated immunity against invading *E. coli* (Yamazaki *et al.*, 1982).

Similar results have been reported in experiments on germ-free rats. When *Bif. longum* was given orally to rats twenty days before administration of a pathogenic *E. coli* strain, no clinical symptoms of infection were noted, but when the rats were given *E. coli* before *Bif. longum*, several of them died within 48 h (Faure *et al.*, 1984). Subsequent studies with germ-free mice monoassociated with *Bif. longum* have indicated an immunological stimulation of the host. The production of anti-*Bif. longum* IgA was accompanied by the production of IgA unrelated to the stimulating antigen; also cell-mediated immunity was activated (Yamazaki *et al.*, 1985; Ueda, 1986).

These data support speculation that ingestion of fermented milks containing large numbers of viable bifidobacteria and lactic acid bacteria may induce an immunological response in humans.

Possible Adverse Effects
Generally, the access of intestinal micro-organisms to soft tissues (which are not their native habitat) has harmful consequences.

Some strains of bifidobacteria have been implicated in possible adverse effects for the host, when found in soft tissues. They usually occur in mixed populations along with facultative anaerobic bacteria (Beerens & Tahon-Castell, 1965; George *et al.*, 1965). Subsequently these strains have been identified as *Bif. dentium* (Scardovi & Crociani, 1974). This species was found in dental caries and other clinical materials, including various abscesses and human pleural fluid (Beerens & Tahon-Castell, 1965; George *et al.*, 1965; Scardovi & Crociani, 1974), thus suggesting its possible harmful potentiality.

SELECTION OF STRAINS OF BIFIDOBACTERIA: TRANSIT IN GASTRO-INTESTINAL TRACT, ADHESION TO EPITHELIAL SURFACE AND SURVIVAL IN CARRIER FOOD

General
In the following section, the selection criteria for human intestinal bacteria as dietary adjuncts are examined. Some complementary details about the selection of strains have been given by Kurmann (1988).

(Current data indicate that a dietary supplementation of intestinal strains of bifidobacteria and lactobacilli for healthy adults does not

TABLE VIII
IMPORTANT CHARACTERISTICS OF CANDIDATE MICRO-ORGANISMS FOR USE AS DIETARY ADJUNCTS

(1) Be a normal inhabitant of the intestinal tract.
(2) Survive the upper digestive tract.
(3) Be capable of surviving and growing in the intestine.
(4) Produce beneficial effects when in the intestinal tract.
(5) Maintain viability and activity in the carrier food before consumption.

Source: Gilliland (1979).

replace other intestinal bacteria, but helps to maintain a proper balance of the resident flora. These organisms have certain characteristics (resistance to unfavourable conditions in the gastro-intestinal tract) which enable them to survive for some time, and possibly to multiply in the

TABLE IX
RESISTANCE TESTS FOR SURVIVING IN THE UPPER DIGESTIVE TRACT

Resistance test method	Micro-organisms checked
(1) *Gastric digestion* in vivo (Mixture with HCl, pepsin and rennet (Breslaw & Kleyn, 1973; Kilara, 1980)	*Lac. acidophilus*—survival (Kilara, 1980) *Propionibacterium freudenreichii*— survival without loss of vitality (Mantere-Alhonen, 1983) *Yoghurt-, buttermilk- and sour milk- cultures*—survival with different digestion times (Kilara, 1980)
(2) *pH* (a) *Human gastric juices* (Gianella *et al.*, 1972)	*Bif. bifidum*—(4 strains), 2 h at pH 2·4 and 6·5. Distinct effect at pH 4·0 and strong action at pH 2 (Lipinska, 1978)
Conditions of the stomach: cultured milk mixed with gastric juice (70:30) Pettersson *et al.*, 1983)	*Lac. acidophilus*—survival (Pettersson *et al.*, 1983; Lindwall & Fonden, 1984) *Yoghurt, sour milk*—Addition of gastric juices with pH 3·48–6·75, no bacteriocidal or bacteriostatic effect observed (Chomakov & Boicheva, 1984)
(b) *Artificial gastric juices* (at pH 3·0, incubation by 37°C—Yakult Honsha Co. Ltd., 1971)	*Lactobacilli—Lac. acidophilus* and *Lac. plantarum* survive 3 h; *Lac. bulgaricus* less resistant, survives 1 h (Yakult Honsha Co. Ltd., 1971)

intestinal tract, before excretion in the faeces. Therefore, regular consumption of fermented milks containing bifidobacteria ensures that they are continually passing through the gut.

In general, candidate organisms for use as dietary adjuncts should have the characteristics mentioned in Table VIII.

There are urease-positive and -negative strains of bifidobacteria. It is important to select only urease-negative strains for therapeutic use (for example, the management of liver diseases). Also, it would be desirable to pay attention to the selection of polysaccharide-producing strains of bifidobacteria, to improve both adhesion to epithelium and the viscosity of products.

Survival of Intestinal Passage: Resistance Tests

The primary barrier to the passage of micro-organisms is the stomach with its gastric acid, and the intensive inhibitory action is mainly related to the concentration of hydrochloric acid. Table IX illustrates some tests for gastric passage and gastric digestion, as well as pH resistance and ability to pass through the stomach. Many references about resistance tests *in vitro* are concerned with micro-organisms other than bifidobacteria, and because of the sparse indications about intestinal resistance, the survey covers those test substances and their concentrations that could be employed in order to check the different micro-organisms of fermented milks.

Table X illustrates some resistance tests for the survival of bacteria in the intestinal lumen (digestive fluids, inhibitory degradations products, deleterious digestive enzymes and administered antibiotics). It is possible that the secretion of bile salts into the duodenum plays some role in limiting bacterial proliferation (Catteau *et al.*, 1971). However, experiments with rats, in which bile salt secretion is prevented, have not resulted in the expected increase in bacterial numbers (Sacquet *et al.*, 1971). Phenols are produced from dietary proteins, and, in particular, from dietary tyrosine, by the gut bacteria, and there are quantitative relationships between protein intake and the amount of certain phenolic compounds in the intestinal lumen (Drasar & Hill, 1974). The resistance to enzymes deleterious to micro-organisms is, at present, limited to the resistance to lysozyme produced by bacteria; lysozyme is capable of lysing certain bacteria. Profound changes may be effected in the normal, human intestinal flora by the administration of various antimicrobial agents. It is, therefore, a sensible precaution to check the resistance of the strains to

TABLE X
RESISTANCE TESTS FOR SURVIVAL IN THE INTESTINES

Resistance test method	Micro-organisms checked
(a) *Digestive fluids: Biliary salts, etc.*	
(1) VL medium with glucose (Catteau *et al.*, 1971)	*Bifidobacteria* —22 strains of *Bif. bifidum*, *Bif. breve*, *Bif. parvulorum*. Bacteriostatic effect with 0·2 or 0·5‰ sodium desoxychlorate (Catteau *et al.*, 1971) —*Bif. bifidum* (4 strains) with 0·5–1 and 3‰ sodium desoxychlorate; an approximately bacteriocidal effect (Lipinska, 1978)
(2) MRS broth and agar (MRS broth with 1·5% agar)	*Lac. acidophilus* Survival in agar with 0·15% oxgall, sensitive to 0·2% oxgall (Brennan *et al.*, 1983, 1986). Agar medium with 0·15% oxgall for enumerating bile resistant lactobacilli (Gilliland, 1979). *Lac. acidophilus* freeze damaged S culture is resistant to 1% and R culture sensitive to 0·6% bile (Klaenhammer & Kleeman, 1981)
(3) Milk or whey peptone medium	*Yoghurt-cultures* Bactericidal effect with 0·25 deoxycholic acid: inferior to 0·25% no action (Bianchi-Salvadori & Ferrari, 1978). Survival at 0·05–0·2% (Bianchi-Salvadori, 1981)
(b) *Inhibitory degradation products: Phenol*	
(1) Milk or MRS broth with addition of 0·1 to 0·8% phenol (Teply, 1984)	*Lac. acidophilus* Able to grow in the presence of 0·5% phenol (Rašić & Kurmann, 1978) *Lac. bulgaricus* Bacteriostatic effect in milk with 0·1–0·5% phenol, and in MRS broth with 0·1 and 0·2% phenol (Kondratenko, 1985). No bacteriostatic action with 0·4% phenol (Maxa & Teply, 1960) *Lactic streptococci* No bacteriostatic effect in milk with 0·3% phenol of a 50% solution in alcohol (Patkul', 1968)

TABLE X—contd.

Resistance test method	Micro-organisms checked
(2) For acid production in milk, see table from Teply (1984) (3) For test of vitality, see Maxa & Teply (1960) and Rašić & Kurmann (1978) (c) *Deleterious digestive enzymes:* *Lysozyme* Nutritive medium with lysozyme addition	*Lac. acidophilus* Isolated from acidophilus milk and dried is sensitive to 100 μg/ml lysozyme (Brennan *et al.*, 1986). Resistance of various animal and human strains was varied and increased in presence of $CaCl_2$. Lactobacilli show a certain resistance
(d) *Therapeutical antibiotics* (1) Serum and Vaccine Institute Warsaw (1976) (2) Various methods	*Bif. bifidum* (4 strains) variable sensitivity (Lipinska, 1978) *Lac. acidophilus* Sensitivity tested by Laanes (1969); Tarabrina & Pinegrin (1979) *Yoghurt starter cultures* Resistance to 35 antibiotics are tested (Sozzi & Smiley, 1980)

therapeutic antibiotics. *Lac. acidophilus* artificially rendered antibiotic-resistant, does, in fact, check the increase of potentially pathogenic staphylococci in the intestine (Gordon *et al.*, 1957).

It would be useful to develop, in the near future, more *in vitro* resistance tests with different inhibition concentrations to simulate the gastro-intestinal environment.

Transit in the Gastro-intestinal Tract

Test for Transit
Transit through the gastro-intestinal tract by fermented milk bacteria implies the ability to survive the passage of the gastro-intestinal environ-

ment in a more or less viable condition. Through analysis of faeces before, during and after transit, the passage of non-intestinal micro-organisms in the gastro-intestinal tract can be easily checked.

Transit through the gastro-intestinal tract may be essential if the ingested food is to have any beneficial effect.

Colonization of the Lumen Content

Colonization of the lumen content means the survival and development of the bacteria for a minimal, albeit limited, period of time. This ensures the maintenance of a certain population, which then disappears several days after discontinuation of ingestion of the fermented food. The regular consumption of fermented milks, in order to maintain a certain population in the lumen content, is needed. This kind of colonization of the lumen content may have a beneficial effect through the depression of harmful enzyme activity, as well as other health-promoting properties.

The limited colonization by lactobacilli, which need a source of carbon, is mainly restricted to the small intestine where the important processes of digestion and absorption of nutrients occur. The bifidobacteria may also interact in the lower part of the small intestine, but in addition, are able to utilize the complex, non-fermentable carbon sources found in the colon. Little is known about the possible colonization of this part of the intestine by transient bifidobacteria.

Adhesion to Epithelial Cell Surface

Some types of indigenous micro-organism adhere to the surfaces of the membranes of the gastro-intestinal epithelial cells. Such organisms can be differentiated into those that induce non-visible changes in the structure of the epithelial cell and those that induce obvious alterations in the membrane of the cell to which they attach (Savage, 1983).

The tests for adhesion to epithelial cell surface of intestinal starter bacteria cannot be performed *in vivo* on humans. The following are examples of tests that can be utilized:

(1) *Animal cells*
    (a) In-vivo adherence tests on different animal cells, e.g. to duodenal brush borders in rats (Cole & Fuller, 1984); adhesion in niches in the mucosal epithelium of germ-free, male mice (Bianchi-Salvadori *et al.*, 1984);
    (b) Shaking a piece of small intestine from a pig or calf in a suspension of the test micro-organism, and examination of the adherent

bacteria, after preparation of slices with a microtome, directly by light microscopy (Mäyrä-Mäkinen et al., 1983).
(2) *Human cells*
 (a) Adherence to the human foetal, intestinal epithelial cell lining (Kleeman & Klaenhammer, 1982).
 (b) Adhesion *in vitro* to human red blood cells (Brilis et al., 1982).

Adhesion is not a necessary requirement for the successful colonization of the intestine, but those micro-organisms that do adhere may have more effect on the physiological functioning of the intestinal tract (Cole & Fuller, 1984).

The use of host-specific culture strains for the production of fermented milk products has been reported to be probably limited to neonatal animals, where the intestinal flora is still in the process of becoming established. In this situation, it may be possible to induce rapid colonization of the gut by desirable organisms. In the adult, however, the flora has been developed to a balanced and stable population, and it seems unlikely that

TABLE XI
THE INFLUENCE OF CARRIER FOOD ON THE VIABILITY OF BIFIDOBACTERIA

Property	*Carrier food parameters*	*Bacterial strains/ starter culture*
pH	Check post-acidification: parameters of manufacture, acidifying capacity of starters; storage temperature; pH not below 4·0–4·2	Aciduric enough
Temperature of storage	Below 5°C	Enough cold resistance
Oxygen contact	Possible packs to reduce oxygen diffusion	Reduced sensitivity to oxygen
Buffer capacity of milk base	Enhanced, e.g. by protein-enriched milk	Not too acid-sensitive strains
Life span	Whole food matrix	Decimal death rate not below 3·5 days
UV-light	Protection in packs	Protection by minimal light exposition and convenient packs
Harmful activity of admixture of bacteria	Cold storage	Resistance, or capacity to destroy H_2O_2, etc.

a new strain would be able to establish itself in a habitat already occupied by the naturally-acquired, indigenous flora (Cole & Fuller, 1984). This consideration may also be valid for human babies.

Survival in Carrier Foods

The carrier food has an important influence on the decimal death rate of bacteria. The carrier food and the selected starter should have the characteristics mentioned in Table XI. The decimal death rate should not be below 3·5 days. Therefore, the bacterial content of each prophylactically important species or strain should not be below 10^8 bifidobacteria/ml after production, and 10^5–10^7 bifidobacteria/ml at the date of consumption of a normal minimum intake of 100 g of fermented milk (Kurmann, 1983), and under the stated storage conditions.

BIFIDOBACTERIA-CONTAINING PRODUCTS

General

Fermented milks containing bifidobacteria are made either using pure strains of these organisms alone, or in combination with other lactic acid bacteria (Table XII). The latter is more common, because the slow acidification of milk by bifidobacteria may be aided by other lactic acid

TABLE XII
FERMENTED MILK PRODUCTS CONTAINING BIFIDOBACTERIA

Product name	Culture organisms
Bifidus milk	*Bif. bifidum* or *Bif. longum*
Bifighurt	*Bif. bifidum*, *Str. thermophilus*
Biogarde	*Bif. bifidum*, *Lac. acidophilus*, *Str. thermophilus*
Biokys	*Bif. bifidum*, *Lac. acidophilus*, *P. acidilactici*
Yoghurt with bifidus bacteria	*Bif. bifidum* (*Bif. longum*), *Str. thermophilus*, *Lac. bulgaricus*
Special yoghurt with bifidus and acidophilus bacteria	*Bif. bifidum* (*Bif. longum*), *Lac. acidophilus*, *Str. thermophilus*, *Lac. bulgaricus*
Cultura	*Bif. bifidum*, *Lac. acidophilus*
Cultura drink	*Bif. bifidum*, *Lac. acidophilus*
Mil-Mil	*Bif. bifidum*, *Bif. breve*, *Lac. acidophilus*
Progurt	*Str. lactis* biovar *diacetilactis*, *Str. lactis* sub-sp. *cremoris*, *Lac. acidophilus* and/or *Bif. bifidum*

Source: Kurmann & Rašić (1988).

bacteria. The first commercial process for the manufacture of fermented milks incorporating bifidobacteria was proposed in 1968 by Schuler-Malyoth et al. (1968).

Bifidobacteria-containing products can be divided, according to their utilization, into three categories: baby foods, bifidus fermented milks and pharmaceutical preparations.

Baby Foods

Bifidobacteria are the predominant intestinal organisms of breast-fed babies. The potentially beneficial roles of these bacteria in the gut include competitive antagonism against pathogens, production of acetic and L(+)-lactic acids and inhibition of nitrate reduction to nitrite; also improved nitrogen retention and weight gain in infants.

Bifidobacteria were used originally in France in 1906 by Tissier, who administered a culture of *Bif. bifidum* to infants suffering from diarrhoea. Subsequently, bifidobacteria were used in the making of baby foods (Mayer, 1948), and bifidus milk containing growth-promoting substances was used in the feeding of infants suffering from nutritional disturbances.

Bifidobacterium bifidum is the species most often used. Either alone, or together with its in-vivo growth-promoting substances, it is said to:

  (a) modify the gut microflora of babies and artificially-fed infants;
  (b) protect against enteric infections or side-effects of antibiotic therapy;
  (c) act as an aid in the therapy of intestinal disorders and enteric infections.

The commercial production of bifidus baby foods began much later. Products incorporating human intestinal strains of bifidobacteria either alone, or together with their growth-promoting substances, include the following preparations:

  (1) The dried formula called 'Lactana-B', containing lactulose and viable *Bif. bifidum*, produced from partially adapted milk (developed in Germany in 1964 (Rašić & Kurmann, 1983)).
  (2) The dried formula product called 'Femilact', containing viable bifidobacteria (developed in Czechoslovakia by Dedicova and Drbohlav (1984)) and made by fermenting heat-treated cream (12% fat) with a mixed culture (2–5%) consisting of *Bif. bifidum*, *Lac. acidophilus* and *Pediococcus acidilactici* (1:0·1:1 ratio) at 30°C to the desired acidity, followed by cooling; heat-treated

vegetable oil, lactose, whey protein and vitamins are added, and the mixture is homogenized and spray-dried. The final product, when reconstituted, contains 0·25% lactic acid and 10^8-10^9/ml viable culture bacteria; the number of viable cells declines by a factor of ten during storage for 2 months.

(3) The liquid formula product called 'Bifiline', containing viable bifidobacteria (developed in USSR by Koroleva et al. (1982)). It is made from a milk formula called 'Malutka' (a baby food) and selected strains of bifidobacteria. The product is made by fermenting the heat-treated, homogenized milk formula with a starter culture (5%) containing 0·5% corn (maize) extract, at 37°C for 8-10 h until coagulation has taken place. After cooling, the final product is reported to contain about 0·60% lactic acid and 10^7-10^9/g viable bifidobacteria.

Fermented Milk Products Containing Bifidobacteria

Fermented milks are prepared with bifidobacteria alone, or in combination with other bacteria. Thermophilic or mesophilic lactic acid-producing bacteria of non-intestinal origin are added in order to help acidification. At the moment, the nutritional–physiological significance of yoghurt-bacteria is better known than that of the mesophilic, lactic acid bacteria. However, the latter have the advantage of producing less lactic acid, which is, of course, inhibitory to intestinal strains of bacteria. Fermented milk products containing bifidobacteria can be divided in the following product groups:

Type 1: Bifidus *Milk*
A fermented product obtained by culturing milk with selected strains of bifidobacteria, most often *Bifidobacterium bifidum*, but sometimes *Bif. longum*. The product is named after the main bacterium (former name— *Lactobacillus bifidus*) used in the fermentation. Bifidobacteria are the predominant intestinal bacteria of breast-fed infants, and a major component of the large intestinal flora in human adults. The naturally-occurring large numbers of these bacteria in the gastro-intestinal tract may be indicative of their importance. As already mentioned, bifidobacteria were first used in 1948 by Mayer (1948) in Germany for the manufacture of baby foods. Subsequently, Schuler-Malyoth and co-workers (1968) suggested the first large-scale commercial process for making fermented milks containing bifidobacteria.

Bifidus milk is produced in small quantities in some European countries, but its consumption is linked to alleged dietetic and therapeutic values rather than to its organoleptic properties.

The product, based on skim-milk, partially skimmed milk or whole milk, is made with viable strains of *Bif. bifidum* or *Bif. longum* originating from the gut (faeces) of healthy humans. Bifidobacteria grow slowly in milk, and produce L(+)-lactic acid and acetic acid in an approximately molar ratio of 2:3; small amounts of formic acid, ethanol and succinic acid are also formed. Their growth and acid production may be improved by using strains that are more acid-tolerant, or by using larger inocula during culturing, and/or adding to the milk such growth stimulating nutrients as yeast extract or autolysate, pepsin-hydrolysed milk, corn extract or whey protein. The mother culture is transferred every 2-3 days to ensure maximum activity, and sterilized skim-milk, with or without added growth-stimulating nutrients, is inoculated with approximately 10% of a culture and incubated at 37-42°C until coagulation and then cooled. Concentrated cultures, with or without growth stimulating nutrients, may be used as bulk starter inocula to avoid sub-culturing in the laboratory.

The manufacture of bifidus milk involves standardization of milk to the desired fat content, increasing the protein content by evaporation, ultrafiltration or addition of skim-milk powder, homogenization and heat treatment (80-120°C for 5-30 min), and tempering. The milk is then inoculated with approximately 10% of a starter culture, mixed well, packed and incubated at 37-42°C until coagulation occurs, followed by cooling. The final product has a pH of 4·3-4·7 and contains 10^8-10^9/ml viable bifidobacteria, whose numbers decline by two log-cycles during refrigerated storage for 1-2 weeks. Bifidus milk can be produced as a stirred product as well (Fig. 4; Rašić & Kurmann, 1983).

Bifidus milk is a food that is claimed to be more easily digested than the milk from which it is made. It has been used as an aid in the therapy of gastro-intestinal disorders, and as a protective measure against imbalance in the gut microflora, as well as in the treatment of liver diseases.

The relatively slow acid production in milk by bifidobacteria, and the unusual flavour of the final product, have led to the development of fermented milk products incorporating other lactic acid bacteria, in addition to bifidobacteria, and the incorporation of fruits. Due to the limited shelf-life of fresh bifidus milk, now there are also available freeze-dried pharmaceutical preparations containing large numbers of viable bifidobacteria, with or without other added lactic acid bacteria.

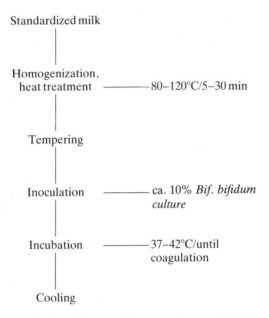

FIG. 4. Flow diagram of the manufacture of *bifidus* milk.

A bifidus-milk with yoghurt flavour is obtained by fermenting whey protein-enriched skim-milk with a culture of *Bif. bifidum* or *Bif. longum*. The technology of this product was developed in the UK in 1982 (Marshall et al., 1982; Gurr et al., 1984) with the purpose of obtaining a dietetic product from specially formulated milk.

The product is made by mixing equal volumes of the retentate of ultra-filtered, sweet whey (concentrated×8) and ultra-filtered skim-milk (concentrated×2), then pasteurizing the mixture at 80°C for 30 min and cooling to 37°C, followed by the additions of threonine (0·1%) and starter (2%). The incubation is at 37°C for 24 h, followed by cooling to 4°C and cold storage.

The final product has about 15% total solids, including 7·3% protein and 1·3% fat, a pH of about 4·7, and an acetaldehyde concentration between 29 and 39 ppm. It contains 10^9/ml viable bifidobacteria, whose numbers decline during storage for 21 days at 4°C to 10^6–10^7 viable cells/ml.

Type 2: Bifidus-Acidophilus-*Milk*
A product called 'Cultura' (Denmark) is made by fermenting

homogenized, heat-treated, protein-enriched whole milk with *Bif. bifidum* and *Lac. acidophilus* at 37°C for about 16 h until the desired acidity is obtained, followed by cooling. It is made by the set method, and the final product contains more than 10^8/ml of *Lac. acidophilus* and more than 10^8/ml of *Bif. bifidum*. It has a characteristic flavour, mild acid taste and firm consistency. 'Cultura' has a shelf life of at least 20 days after production, and is sold in plastic containers of 150 ml. By fermenting partially skimmed milk, a drink is obtained (Hansen, 1985).

A similar drink called 'Mil-Mil' (Japan) is made by fermenting the heat-treated milk with a starter which incorporates *Bif. bifidum*, *Bif. breve* and *Lac. acidophilus*. It is sweetened with small amounts of glucose or fructose, and is coloured with carrot juice (Anon, 1978).

Type 3: Bifidus-Thermophilus-*Milk*
'Bifighurt' (Germany) is made by fermenting heat-treated milk with 6% of a starter consisting of *Bif. longum* and *Str. thermophilus* at 42°C for about 4 h. The product has a pH of about 4·7, mild taste and may contain 10^7/ml *Bif. longum* as well as a large number of *Str. thermophilus* (Klupsch, 1986).

Type 4: Bifidus-Acidophilus-Thermophilus-*Milk*
'Biogarde' (Biogarde Company) is obtained by fermenting milk with a starter of *Bif. bifidum*, *Lac. acidophilus* and *Str. thermophilus* (Schuler-Malyoth *et al.*, 1968). The process for creating the bulk starter is as follows: the special nutrient medium for Biogarde culture (1·5%) is dissolved in water; this solution is added to the bulk starter milk; heated to 90°C for 10 min; cooled to 42°C; inoculated with liquid, freeze-dried or deep-frozen Biogarde culture; incubated for 4·5–6·5 h at 41–42°C depending on the quantity of starter inoculum; cooled to, and stored at, approximately 8°C. The manufacture of 'Biogarde' involves standardization of the milk, homogenization, heat treatment (90°C for 10 min or 95°C for 5 min), tempering and inoculation with a bulk starter (6%), mixing well, packaging and incubating at 42°C for about 3·5 h until coagulation, then cooling. 'Biogarde' is also used in the preparation of various other products, including buttermilk, sauces, muesli (breakfast cereal) beverages, fresh cheeses, ice-cream and beverages.

The final product reportedly contains 0·85–0·90% L(+)-lactic acid, about 10^7–10^8 cells of *Lac. acidophilus* per ml and 10^6–10^7 *Bif. bifidum* cells per ml, in addition to a large number of *Str. thermophilus* cells (Klupsch, 1984, 1985).

Type 5: Bifidus-Acidophilus-Pediococcus *Milk*
'Biokys' (Czechoslovakia) is obtained by fermenting milk with a mixed culture of bifidobacteria, *acidophilus* bacteria and *Pediococcus acidilactici*. The first two cultures are human intestinal strains, while *Ped. acidilactici* is incorporated to help with the acidification.

The technology of 'Biokys' was developed in Czechoslovakia, where this product is commercially produced as a health product. It is made using standardized milk (15% total solids including 3·5% fat), which is homogenized and heat-treated, then fermented with 2–5% of a starter consisting of *Bif. bifidum. Lac. acidophilus* and *Ped. acidilactici* (1:0·1:1 ratio) at 30–31°C to the desired acidity, followed by stirring and cooling. The product has a sour cream-like viscosity, and a clean and mild acid taste (Hylmar, 1978).

Type 6: Bifidus-*Yoghurt* or Bifidus-Acidophilus-*Yoghurt*
The combination of yoghurt culture bacteria with bifidobacteria or, additionally, *Lac. acidophilus* has led to the development of products which have a characteristic flavour (Mülhens & Stamer, 1969; Rašić & Kurmann, 1978).

There are two types of such yoghurts containing bifidobacteria:

(6a) The first sub-group contains only bifidobacteria and yoghurt organisms. It is produced in Germany, USA, Japan, France and in several other countries. The product is made either by a simultaneous fermentation of the standardized, heat-treated, homogenized milk with a starter (5%) consisting of a yoghurt culture and bifidobacteria (*Bif. bifidum* or *Bif. longum*) at 40–42°C for 3–4 h until coagulation, followed by cooling, or by mixing into cultured yoghurt, at the desired ratio, a separately cultured bifidus milk followed by stirring and cooling. The former method may possibly ensure better viability of the bifidobacteria (Rašić, 1987). Recent examples of this type of product are 'B*A' (denotes bifidus active), with a human strain of *Bif. longum* and a yoghurt microflora, produced in France and 'Real Active' marketed in the UK.

(6b) The second sub-group of products is usually made by a simultaneous fermentation of the heat-treated milk with cultures of *Bif. bifidum* and/or *Bif. longum, Lac. acidophilus* and the yoghurt micro-organisms. The product has a characteristic, mild acid flavour, which may be masked or modified in fruit-flavoured varieties. A more recent example of this type of product is 'Ofilus', France.

Type 7: Bifidus-Acidophilus-*Mesophilic* Streptococcus-*Milk*
A protein-enriched product called 'Progurt' (Schacht & Syrazynski, 1975) is prepared by fermenting pasteurized skim-milk with 1–3% of a mixed culture of *Str. lactis* subsp. *diacetylactis* and *Str. lactis* subsp. *cremoris* (ratio 1:1) until 0·7–0·8% lactic acid is produced. The product is then subjected to a partial whey separation, the addition of cream and 0·5–1·0% of *Lac. acidophilus* and/or *Bif. bifidum* cultures, homogenization and cooling. The product has a 5% fat, 6% protein, 3% lactose content, and a pH of 4·4–4·5. A product made in France, called 'Ofilus Double Douceur' contains *Bif. bifidum, Lac. acidophilus, Str. lactis* subsp. *lactis* and *Str. lactis* subsp. *cremoris*.

Pharmaceutical Preparations
Freeze-dried, pharmaceutical preparations containing bifidobacteria alone, or in combination with other organisms, have been described in detail by Rašić and Kurmann (1983). They are prepared by the pharmaceutical and food industries, and contain autochthonous intestinal micro-organisms. In general, these preparations are utilized for the therapy of gastro-intestinal disturbances (diarrhoea, side-effects of antibiotic and radiation therapy), and as special preparations for certain liver diseases.

Some examples of pharmaceutical preparations are Bifidogene (1977, France); Bifider (Japan, *Bif. bifidum*); Infloran Berna (*Lac. acidophilus, Bif. infantis*, Switzerland); Euga-Lein, Eugalan Topfer forte, and Lactopriv (from Topfer GmbH, Germany, with bifidobacteria strains); Preparation J (containing *Bif. bifidum*, Yugoslavia) and Lyobifidus containing *Bif. bifidum* (Tissier, France) (Răsić & Kurmann, 1983).

CONCLUSIONS AND SUMMARY

Fermented fresh milk products containing selected human intestinal strains of bifidobacteria may be regarded as foods possessing health-promoting as well as prophylactic, probiotic and therapeutic properties.

Health-promoting properties: the nutritional-physiological value of fermented milk products may be based on the effect of factors such as: (a) the well-known properties of milk as a raw material; (b) the products of fermentation (organic acids, peptides, enzymes, etc.); (c) the bioactivity of live bifidobacteria in the intestines.

Prophylactic properties: the bioactivity of bifidobacteria may be based on the following effects: (a) competitive antagonisms against invading pathogens (competition for nutrients and for attachment sites to epithelial surfaces); (b) production of organic acids (acetic, lactic) and possibly, by some strains, other antimicrobial substances; (c) lowering the activity of some harmful bacterial enzymes and, consequently, decreasing the formation of harmful products (amines, ammonia, nitrosamines, indican, etc.); (d) probiotic effect—contribution to the balance of the intestinal microflora.

Therapeutic properties: some evidence suggests that large numbers of bifidobacteria in the intestine have beneficial effects in cases such as: (a) treatment of enteric infections; (b) treatment of intestinal disturbances, e.g. side-effects of antibiotic therapy or irradiation therapy; (c) restoration of the disturbed intestinal microflora due to other stress conditions (cosmonauts, etc.); and (d) management of liver diseases.

Factors Influencing the Activity of Bifidobacteria in the Intestines
The beneficial effects of ingested bifidobacteria may be obtained: (a) if selected strains are used; (b) if large numbers of viable cells (10^8–10^9 cells/day) are introduced; (c) if they survive the gastro-intestinal transit and preferably adhere to epithelial surfaces and grow; (d) if a fermentable carbohydrate is available to the cells in the intestine; (e) if the carrier food (type of food and its biochemical characteristics, type of package, storage condition) enables a high performance of selected strains.

The mechanisms of the beneficial effects of bifidobacteria are still insufficiently known, and further research and development are needed.

REFERENCES

Anand, S. K., Srinivasan, R. A. & Rao, L. K. (1984). *Cultured Dairy Products Journal*, November, 6–8.
Anand, S. K., Srinivasan, R. A. & Rao, L. K. (1985). *Cultured Dairy Products Journal*, February, 21–3.
Anon. (1978). *Packaging*, **49**, 37.
Bamberg, H. (1966). *Med. Welt*, **39**, 2086.
Beerens, H., Romond, C. & Neut, C. (1980). *American Journal Clinical Nutrit.*, **33**, 2434–9.
Beerens, H. & Tahon-Castell, M. M. (1965). *Infections humaines à bactéries anaérobies non toxigènes*. Presse Académiques Européennes, Bruxelles.

Benno, J. & Mitsuoka, T. (1986). *Bifidobacteria Microflora,* **5,** 13–25.
Bianchi-Salvadori, B. (1981). Symposium international sur les effets nutritionels de la flore digestive, Texte des exposés. Syndifrais, Paris, France, pp. 66–87.
Bianchi-Salvadori, B. & Ferrari, A. (1978). *Latte,* **3,** 249–55.
Bianchi-Salvadori, B., Camaschella, P. & Bazzigaluppi, E. (1984). *Milchwissenschaft,* **39,** 387–91.
Biavati, B., Castagnoli, P. & Trovatelli, L. D. (1986). *Microbiologica (Bologna),* **9,** 39–46.
Biogarde Company; S.BI SANOFI Bio-Industries GmbH, Kanzlerstrasse 6, D-4000 Düsseldorf 30, West Germany.
Branca, G., Salvaggio, E., Manzara, St. & Paone, S. M. (1979). *Clin. Terapia,* **90,** 565–78.
Braun, O. H. (1971). In *Moderne Aspekte der künstlichen Säuglingsernährung,* ed. H. Berger. G. Thieme, Stuttgart, pp. 139–46.
Brennan, M., Wanismail, B. & Ray, B. (1983). *Journal Food Protect.,* **46,** 887–92.
Brennan, M., Wanismail, B., Johnson, M. C. & Ray, B. (1986). *Journal Food Protect.,* **49,** 47–53.
Breslaw, E. S. & Kleyn, D. H. (1973). *Food Science,* **38,** 1016–21.
Brezina, P., Horky, J. & Plockova, M. (1983). *Prumysl Potravin,* **34,** 538–41.
Brilis, V. I., Brilene, T. A., Lentsner, Kh.P. & Lentsner, A. A. (1982). *Zhurnal Mikrobiologii, Epidemiologii i Immunologii,* **9,** 75–8.
Buchanan, R. E. & Gibbons, N. E. (1974). *Bergey's Manual of Determinative Bacteriology,* 8th Edn. Williams & Wilkins, Baltimore.
Bullen, C. L., Rogers, H. Y. & Ligh, L. (1972). *Brit. Med. Journal,* **1,** 69.
Bullen, C. L. & Tearle, P. V. (1976). *J. Med. Microbiol.,* **9,** 335–44.
Bullen, C. L., Tearle, P. V. & Willis, A. (1976). *J. Med. Microbiol.,* **9,** 325–33.
Catteau, M., Henry, M. & Beerens, H. (1971). *Ann. Inst. Pasteur, Lille,* **22,** 201–5.
Chomakov, Kh. & Boicheva, S. (1984). *Zhivotnov dni Nauki,* **21,** 91–6.
Chung, K.Ch. & Goepfert, J. M. (1970). *J. Food Sci.,* **35,** 326–8.
Cole, C. B. & Fuller, R. (1984). *J. Appl. Bact.,* **56,** 495–8.
Conge, G. A., Gonache, P., Desormeau, J. P., Loisillier, F. & Lemonier, D. (1980). *Reprod. Nutr. Develop.,* **20,** 929–38.
Dedicova, L. & Drbohlav, J. (1984). FIL/IDF-Bulletin Doc. 179, posters p. VI.
Dehnert, J. (1957). *Zbl. Bakt. (I Abt. Orig.),* **169,** 66–83.
Dolezalek, J. (1979). *Prumysl Potravin,* **30,** 684–5.
Dorofeichuk, V. G., Volkov, A. I., Kulik, N. N., Karaseva, G. N. & Zimina, N. S. (1983). *Voprosy Pitaniya,* **6,** 30–3.
Drasar, B. S. & Hill, M. J. (1974). *Human Intestinal Flora.* Academic Press, London, New York.
Ervolder, T. M., Gudkov, A. V., Gudkov, S. A., Dushenin, N. V. & Trubnikov, N. K. (1984). *Mol. Prom,* **8,** 18–20.
Faure, J. C., Schellenberg, D. A., Bexter, A. & Wuerzner, H. P. (1984). *Zeitschr. für Ernährungswiss,* **23,** 41–51.
Fomon, S. J. (1974). *Infant Nutrition,* 2nd Edn. W. B. Saunders Co., Philadelphia.
Frisell, E. (1951). *Acta paediat., Stockholm* 40 Suppl., **80,** 1–123.

Georg, L. K., Roberstad, G. W., Brinkman, S. A. & Hicklin, M. D. (1965). *J. Infect. Dis.*, **115**, 88–9.
Gianella, R. A., Broitman, S. A. & Zamcheck, N. (1972). *Gut*, **13**, 251–6.
Gilliland, S. E. (1979). *J. Food Protect.*, **42**, 164–7.
Gorbach, S. L., Nahas, L., Lerner, P. I. & Weinstein, L. (1967). *Gastroenterology*, **53**, 845–55.
Gordon, D., Macrae, J. & Wheater, D. M. (1957). *Dairy. Sci. Abstr.*, **19**, 567.
Grütte, F. K. & Müller-Beuthow, W. (1980). In *Gastrointestinale Microflora des Menschen*, ed. H. Bernhardt & M. Knoke. Johann Ambrosius Barth, Leipzig, pp. 39–43.
Gurr, M. I., Marshall, V. M. E. & Fuller, R. (1984). *FIL/IDF Bulletin*, **179**, 54–9.
Haenel, H. (1970). *Am. J. Clin. Nutrit.*, **23**, 1433–9.
Haenel, H., Müller-Beuthow, W. & Grütte, F. K. (1970). *Typen. Zbl. Bakt. (I Abt. Orig.)*, **215**, 333–47.
Haller, J. & Kräubig, H. (1960). *Strahlenther.*, **113**, 272.
Hambraeus, L. (1978). Nutritive value of milk protein. 20th Int. Dairy Congr., Paris. Conferences 19 ST.
Hansen, R. (1985). *Nord. Mejeritidsskr.*, **3**, 79–83.
Hoffmann, K. (1966). *Bakterielle Besiedlung des Menschlichen Darmes*. Dr. A. Hüthig, Heidelberg.
Horečny, K. (1964). *Čsl. Paediat.*, **19**, 673–80.
Hylmar, B. (1978). *Prumysl. Potravin*, **29**, 99–100.
Inoue, K. & Nagayama, T. (1970). *Acta paediat. Jap. (Overseas)*, **12**, 15–20.
Jelliffe, D. B. & Jelliffe, E. F. (1981). *Lancet*, 22 August 1981, 419.
Kageyama, T., Romoda, T. & Nakano, J. (1984). *Bifidobacteria Microflora*, **3**, 29–33.
Kaloud, H. & Stögmann, W. (1968). *Arch. Kinderheilk.*, **177**, 29–35.
Kilara, A. (1980). *J. Dairy Sci.*, **63** supplement 1, 52.
Klaenhammer, T. R. & Kleeman, E. G. (1981). *Appl. Env. Microbiol.*, **41**, 1461–7.
Klaus, M. H. & Dias-Rossello, J. (1980). *J. Paediat. Rev.*, **1**, 289–93. Cited in Jelliffe, D. B. & Jelliffe, E. F. (1981).
Kleeman, E. G. & Klaenhammer, T. R. (1982). *J. Dairy Sci.*, **65**, 2063–9.
Klupsch, H. J. (1984). *Saure Milcherzeugnisse, Milchmischgetränke und Desserts*, Verlag Th. Mann, 4250 Gelsenkirchen-Buer.
Klupsch, H. J. (1985). *South African J. Dairy Technol.*, **17**, 153–6.
Klupsch, H. J. (1986). FRG Patent Application 3120 505 C2 of 24 April 1986.
Kohwi, Y., Hashimoto, J. & Tamura, Z. (1982). *Bifidobacteria Microflora*, **1**, 61–8.
Koizumi, T., Mitsuya, N., Nukuta, T., Fujita, S., Ishizu, H. & Yoshioka, H. (1980). *Sogo Rinsho*, **29**, 2473–8.
Kondratenko, M.St. (1985). *Dtsche. Molk.-Zeitung*, **105**, 710–18.
Koroleva, N. S., Semenikhina, V. E., Ovanova, I. N., Oleneva, I. V. & Sundukova, K. B. (1982). *Moloch. Prom.*, **6**, 17–20.
Korshunov, V. M., Sinitsyua, N. A., Ginodman, G. A. & Pinegin, B. V. (1985). *ZH. Mikrobiol. Epidemiol. Immunobiol.*, **9**, 20–5.
Kurmann, J. A. (1983). *Dtsch. Milchwirtschaft*, **34**, 658.
Kurmann, J. A. (1986). *Dtsch. Molk.-Ztg.*, **107**, 1470–8.

Kurmann, J. A. (1988). *Fermented Milks. Science and Technology, FIL/IDF Bull.*, **227**, 41–55.
Kurmann, J. A. & Rašić, J.Lj. (1988). In *Fermented Milks. Science and Technology, FIL/IDF Bull.*, **227**, 101–9.
Kurmann, J. A. (1989). SMK-Schrift No. 2, Verlag Schweizerische Milchkommission, Bern., 50–61.
Laanes, S.Kh. (1969). *Antibiotiki*, **14**, 426–9.
Lemonnier, D. (1984). *Fermented Milks and Health, FIL/IDF Bull.*, **179**, 60–6.
Leperchy, E. & Frottier, J. (1965). *Sem. Hôp. Paris*, **41**, 221–7.
Levesque, J., Aicardi & Gautier (1959). *Sem. Hôp. Paris*, **35**, 262–7. Cited in DSA, **24**, 141 (1962).
Lindwall, S. & Fonden, R. (1984). *Fermented Milks and Health, FIL/IDF Doc.*, **179**, 21.
Lipinska, E. (1978). *20th Int. Dairy Congr. Paris*, E, 528–9.
Mada, M. (1981). *Jap. Journal Dairy Food Sci.*, **30**, 205–17.
Mantere-Alhonen, S. (1983). *Meijeritieteellinen Aikakauskirja*, **41**, 19–23.
Marshall, V. M., Cole, W. M. & Mabbitt, L. A. (1982). *J. Soc. Dairy Techn.*, **35**, 143–4.
Mata, L. Y., Carillo, C. & Villatoro, M. (1969). *Appl. Microbiology*, **17**, 596–602.
Maxa, V. & Teply, M. (1960). *Vyzkum. Prumysl. Potravin*, **6**, 217–37.
Mayer, J. B. (1948). *Z. Kinderheilkunde*, **65**, 319–45.
Mayer, J. B. (1966). *Mschr. Kinderheilk.*, **114**, 67–73.
Mäyrä-Mäkinen, A., Manninen, M. & Gyllenberg, H. (1983). *J. Appl. Bact.*, **55**, 241–5.
Minigawa, K. (1970). *Acta paediat. Jap. (Domestic)*, **74**, 761–7.
Mitsuoka, T. (1969). *Zbl. Bakt. (I. Abt. Orig.)*, **210**, 52–64.
Mitsuoka, T. (1984). *Bifidobacteria Microflora*, **3**, 11–28.
Mizutani, T. & Mitsuoka, T. (1980). *Cancer Lett.*, **11**, 89–95.
Momose, H., Igarachi, M., Kawashima, T. & Kuboyama, M. (1982). *21st Int. Dairy Congr. Moscow*, **1**(2), 348.
Moreau, M. C., Ducluzeau, R., Guy-Grand, D. & Muller, M. C. (1978). *Infect. Immun.*, **21**, 532–9.
Mülhens, K. & Stamer, H. (1969). *Milchwissenschaft*, **24**, 25–8.
Mutai, M. & Tanaka, R. (1987). *Bifidobacteria Microflora*, **6**, 23–41.
Müting, D. (1977). *Leber, Magen, Darm*, **7**, 256–62.
Müting, D. & Ordnung, W. (1976). *Therapie Woche*, **26**, 1590.
Müting, D., Reikowski, H., Escherich, W., Klein, Ch. & Doenecke, D. (1968) *Dtsche med. Wschr.*, **93**, 1313–19.
Nakaya, R. (1984). *Bifidobacteria Microflora*, **3**, 3–9.
Neumeister, K. & Schmidt, W. (1963). *Med. Klin.*, **58**, 842–4.
Okamura, N., Nakaya, R., Yokota, H., Yanai, N. & Kawashima, T. (1986). *Bifidobacteria Microflora*, **5**, 51–5.
Ordnung, W. & Müting, D. (1973). *Aerztliche Praxis*, **25**, 85–8.
Orla-Jensen, S., Olsen, E. & Geill, T. (1945). *Biol. Skr. K. danske vidensk. Selsk.*, **3**, 3–38.
Pahwa, A. & Mathur, B. N. (1982). *Indian Journal Nutrit. and Diet.*, **19**, 267–72.
Patkul', G. M. (1968). *Moloch. Prom.*, **29**, 6–9.

Petterson, L., Graf, W. & Sewelin, U. (1983). In *Symposium of the Swedish Nutrition Foundation. XV. Nutrition and the Intestinal Flora*, ed. Bo Hallgren. Almquist & Wiksell International, Stockholm, Sweden.
Petuely, F. & Linder, G. (1965). *Zbl. Bakt. (I. Abt. Orig.)*, **195**, 347–84.
Poupard, J. A., Husain, J. & Norris, R. F. (1973). *Bact. Rev.*, **37**, 136–65.
Rašić, J.Lj. (1983). *North Europ. Dairy Journal*, **4**, 80–8.
Rašić, J.Lj. (1984). *Fermented Milks and Health, FIL/IDF Bull.*, **179**, 27–31.
Rašić, J.Lj. (1987a). *Cult. Dairy Prod. Journal*, August, 6–9.
Rašić, J.Lj. (1987b). *Proceedings of the 22nd Int. Dairy Congr.* D. Reidel Publishing Company, Dordrecht, Holland, pp. 673–82.
Rašić, J.Lj. & Kurmann, J. A. (1978). *Yoghurt. Scientific Grounds, Technology, Manufacture and Preparations.* Technical Dairy Publishing House, Copenhagen, Denmark.
Rašić, J.Lj. & Kurmann, J. A. (1983). *Bifidobacteria and their Role. Microbiological, Nutritional-Physiological, Medical and Technological Aspects and Bibliography.* Birkhäuser Verlag, Basel, Switzerland.
Reuter, G. (1963). *Zbl. Bakt (I. Abt. Orig.)*, **191**, 486–507.
Roberts, A. K., Biervliet, J. P. van & Harzer, G. (1985). In *Composition and Physiological Properties of Human Milk, Proceedings of the International Workshop,* Kiel 1985, ed. J. Schaub. Elsevier, Amsterdam, Netherlands, pp. 259–69.
Rowland, I. R. & Grasso, P. (1975). *Appl. Microbiol.*, **29**, 7–12.
Sacquet, E., Raibaud, P. & Garnier, J. (1971). *Ann. Inst. Past.*, **120**, 501–24.
Sahota, S. S., Brawley, P. M. & Menzies, I. S. (1982). *J. Gen. Microbiol.*, **128**, 319–25.
Sato, J., Mochizuki, K. & Homma, N. (1982). *Bifidobacteria Microflora*, **1**, 51–4.
Savage, O. C. (1977). In *Microbial Ecology of the Gut*, ed. R. T. Y. Clarke & T. Bauchop. Academic Press, London, pp. 277–310.
Savage, D. C. (1983). In *Human Intestinal Microflora in Health and Disease*, ed. D. J. Hentges. Academic Press, London, pp. 55–74.
Scardovi, V. (1986). In *Bergey's Manual of Systematic Bacteriology*. Volume 2, ed. P. H. A. Sneath, N. S. Mair, E. Sharpe & J. G. Holt. Williams & Wilkins, Baltimore, pp. 1418–34.
Scardovi, V. & Crociani, F. (1974). *Int. Journal Syst. Bact.*, **24**, 6–20.
Schacht, E. & Syranzinski, A. (1975). *Ind. Lechera*, **646**, 9–11.
Schneegans, E., Haarscher, A., Lutz, A., Lévy-Silage, J. & Schmittbühl, J. (1966). *Sem. Hôp. Paris*, **42**, 457–62.
Schuler-Malyoth, R., Ruppert, A. & Müller, F. (1968). *Milchwissenschaft*, **23**, 356–60.
Serum and Vaccine Institute, Warsaw (1976). Cited by E. Lipinska (1979). *Acta Alimentaria Polonica*, **5**, 359–63.
Sgorbati, B., Scardovi, V. & Leblanc, D. J. (1982). *J. Gen. Microbiol.*, **128**, 2121–31.
Solomonov, K., Kolev, A., Vitkov, V., Nikolov, Z., Tosheva, J. & Prvulov, B. (1984). *Zhivotnovodni Nauki*, **21**, 40–6.
Sozzi, T. & Smiley, M. B. (1980). *Appl. Environm. Microbiol.*, **40**, 862–5.
Stark, P. L. & Lee, A. (1982). *J. Med. Microbiol.*, **15**, 189–203.
Stenger, K. & Wolf, H. (1962). *Arch. Kinderheilk. Beiheft Nor.*, **46**.

HEALTH POTENTIAL OF PRODUCTS CONTAINING BIFIDOBACTERIA

Takano, R., Arai, K., Murota, I., Hayakawa, K., Mizutani, T. & Mitsuoka, T. (1985). *Bifidobacteria Microflora*, **4**, 31–7.

Tamura, Z. (1983). *Bifidobacteria Microflora*, **2**, 3–16.

Tarabrina, N. P. & Pinegrin, V. B. (1979). *Zhurnal Mikrobiologii Epidemiologii i Immunologii*, **7**, 88–92.

Tasovac, B. (1964). *Ann paediat.*, **40**, 1315–22.

Teply, M. (1984). *Cisté Mlékarské Kultury*, Praha, SNTL Nakladatelstvi, Technikcké Litertury.

Timoshko, N. A., Vilshanskaya, F. L., Pospelova, V. V. & Rakhimova, N. G. (1979). *Z. Mikrobiol. Epidemiol. i. Immunobiol.*, **7**, 92–6.

Tissier, H. (1900). Recherches sur la flore intestinale des nourrissons (etat normal et pathologiques). Thesis Université Paris, George Carré et C. Naud, Paris.

Tohyama, T., Tanaka, R., Kobayashi, Y. & Mutai, M. (1982). *Bifidobacteria Microflora*, **1**, 45–50.

Töpfer GmbH. (1980). Prospectus. Töpfer GmbH, Dietmannsried, Allgäu, German Fed. Republic.

Ueda, K. (1986). *Bifidobacteria Microflora*, **5**, 67–72.

Werner, H. (1966). *J. Appl. Bact.*, **29**, 138–46.

Yakult Honsha Co. Ltd. (1971). The summary of Reports Yakult. Yakult Honsha Co. Ltd. 1-1-19, Higshi Shinbashi, Minato-ku, Tokyo, Japan and *Intestinal flora of microorganisms and health*. Prospectus.

Yamazaki, Sh., Kamimura, H., Momose, H., Kawashima, T. & Ueda, K. (1982). *Bifidobacteria Microflora*, **1**, 55–9.

Yamazaki, Sh., Machii, K., Tsuyuki, S., Momose, H., Kawashima, T. & Ueda, K. (1985). *Immunology*, **56**, 43–50.

Yuhara, T., Isojima, S., Tsuchiya, F. & Mitsuoka, T. (1983). *Bifidobacteria Microflora*, **2**, 33–9.

Chapter 7

PRODUCTS PREPARED WITH LACTIC ACID BACTERIA AND YEASTS

N. S. KOROLEVA

International Dairy Federation, Moscow, USSR

Fermented milk products have long been an important component of the national diet in the countries of Europe, Asia and Africa. They were prepared at home by traditional methods for each country, as a rule, using a natural starter. Such products include: prostokvasha, varenec (Central Russia); ryazhenka (Ukraine); viscous milk—teatta (Scandinavia); yoghurt (Turkey); kiselo mlyako (Bulgaria); Leben (Egypt, Syria, Lebanon); matsun, matsoni, ayran, kefir (Caucasus); kuruga, katyk, chal (Central Asia and Kazakhstan); kumys (Kazakhstan, Mongolia, Lower and Middle Volga region); dahi (India); and mast (Iran), etc.

A distinctive feature of these products is that, in addition to the lactic acid fermentation induced by the activity of lactic acid bacteria, an alcoholic fermentation due to yeasts present, and sometimes acetic acid production by acetic acid bacteria, often takes place.

When industrial production of fermented milk products with pure cultures began, only a few of the great number of national products were selected. Most of the products lost their original natural microflora. Among these are yoghurt, widespread in the countries of Western Europe, America and Australia, kiselo mlyako in Bulgaria, matsun and matsoni in the Caucasus and prostokvasha, varenec and ryazhenka in the USSR; all of which are prepared with pure cultures of *Str. thermophilus* and *Lac. delbrueckii* sub-sp. *bulgaricus* (*Lac. bulgaricus*).

However, in some fermented products manufactured in Eastern Europe and Asia, the original natural microflora was preserved, including yeasts and, sometimes, acetic acid bacteria. Such products are kumys and kefir. The products are considerably different from those prepared from pure cultures of lactic acid bacteria in technical characteristics, in microbiological, physical, chemical and organoleptical features, and also in the therapeutic effect on humans.

Along with the well-known kefir and kumys, some other products have been developed in the USSR which are prepared with lactic acid bacteria and yeasts. These are acidophilin and acidophilus-yeast milk.

KEFIR

The History of Kefir Production in the USSR
The history of kefir dates back to ancient times. The motherland of the product is considered to be the northern slopes of the Caucasian Chain. Since immemorial times, caucasian inhabitants have learnt to make a refreshing drink from cow and goat milk using kefir grains as a starter. Kefir was prepared in wine-skins (leather sacks); during the day they were exposed to the sunlight, at night they were taken into the house and hung near the door. Everyone who went in or out had to push the sack with his/her foot in order to mix the contents. As the kefir was consumed, some fresh milk was added. The finished product was characterized by high acidity and, depending on holding time, higher or lower CO_2 and alcohol contents. At the end of the last century, after the publication of works on kefir properties, kefir grains were brought from the Caucasus with great difficulty and, in 1908, its production in small quantities (not more than 500 litres per day) was organized. As a starter culture, kefir grain 'washes' were used (that is milk fermented by kefir grains and then separated from them—so called 'grain starter'). The fermented milk was filled into bottles, tightly capped and shaken several times while ripening. The finished product was highly carbonated and contained up to 2% alcohol.

The industrial manufacture of kefir in the USSR started in the 1930s. First, the set method was used for kefir production. Inoculated milk was filled into bottles, fermented in thermostatically-controlled areas until a strong coagulum had formed, then cooled in refrigeration chambers. The quality of such products was worse than that obtained by traditional technology.

At the end of the 1950s, a group of specialists from the All-Union Dairy Research Institute headed by M. G. Demurov developed a method of kefir manufacture that provided a product with properties close to that of the traditional, national product, known as 'stirred' kefir. According to this method, all the processes—fermentation, coagulum formation, agitation, ripening and cooling—take place in one large vessel with a cooling jacket. At present, it is the main method of kefir manufacture in the USSR. Kefir produced by this method is a fermented drink with a

PRODUCTS PREPARED WITH LACTIC ACID BACTERIA AND YEASTS 161

typical, refreshing, slightly sharp taste and aroma. Its alcohol content is minimal (traces), and no gas is formed.

Starter
Kefir starter is prepared by growing kefir grains in milk. It is the only natural starter used in industrial kefir production at the present time. Up until now, no-one knows where and how kefir grains appeared. Among the people living in the Caucasian mountains, there exists a legend that kefir grains were given to the Orthodox people by Mahomet, who indicated how to use them. Mahomet strictly forbade them to give away the secret of kefir preparation to other peoples, or pass to anybody kefir grains—'Mahomet grains'—because they can lose their 'magic strength'. The legend explains the fact that kefir grains and the method for kefir preparation have been kept secret and surrounded by mystery for so long.

Kefir grains are characterized by an irregular form, folded or uneven surface, white or yellowish colour, elastic consistency and specific acid taste. Their diameter may be in the range of 1–6 mm or more, depending on the intensity of the agitation during their growth. The more intense the agitation, the smaller the kefir grains in size, and the larger the surface area in contact with the milk, the higher the activity of the microflora.

Active kefir grains float on the surface of the milk. In spite of the efforts undertaken, no-one has succeeded in obtaining a new kefir grain, possessing the structure and properties typical of the original, from a mixture of separate micro-organisms comprising the kefir grain microflora. In practice, new kefir grains are obtained by the growth and proliferation of existing grains.

The study of kefir grains by microscopy shows closely interwoven rod-like threads forming the basis of the grain and keeping the other micro-organisms together. Studies of this basic microbe were described in many publications beginning from the 1930s. Pick, in 1933 (cited from Voitkevich, 1948), separated a thick rod from kefir grains that contained glycogen inclusions. Mutual development with yeasts caused a 'ball' formation. The rod did not grow on dense nutritive media in pure culture, and its proliferation was observed only on growing with yeasts. The microbe is supposed to be a spore-forming rod, which as a result of its development together with yeasts and other components of kefir grains, lost its spore-forming properties and acquired the ability to grow at low temperatures and form lactic acid from lactose. Some scientists regard the basic rod as the lactic acid microbe.

Using the special media KPL and MRS, Toba et al. (1986) separated from kefir grains a homofermentative strain of *Lactobacillus,* which produces a polysaccharide identical to the one extracted from kefir grains. The micro-organism may be involved in kefir grain soma formation. Non-lactose fermenting yeasts (*Torulaspora* type) are most tightly connected with the basic rod. The micro-organisms remain in the kefir grain after careful washing of the surface with water and ethyl spirit (Bukanova, 1955). The surface microflora of the kefir grain includes yeasts fermenting lactose (*Saccharomyces* sp., *Candida* type), mesophilic homo- and heterofermentative streptococci, mesophilic and thermophilic lactobacilli, and acetic acid bacteria.

When kefir grains are introduced into milk, it is mainly the surface microflora that participates in the formation of the kefir starter microflora. The starter microflora includes five basic, functional groups that have a noticeable effect on starter activity and kefir quality:

(a) Mesophilic, homofermentative, lactic acid streptococci (*Str. lactis, Str. lactis* sub-sp. *cremoris*), are the most active part of the kefir starter, surface microflora (Koroleva, 1966, 1975, 1988), and provide a rapid increase in acidity during the first hours of fermentation. At higher acidities, they are inhibited.

(b) Lactobacilli—the obligate heterofermentative *Lac. brevis* (Syn. *Betabacterium,* Arla, Jensen), and the facultative heterofermentative *Lac. casei* sub-sp. *rhamnosus* (Syn. *Streptobacterium*) are most common in kefir starters. The taxonomic position of the obligate heterofermentative lactobacilli present in kefir grains was recently revised (Kandler & Kunath, 1983) using phenotypic and genetic characteristics. All the strains studied were different from *Lac. brevis* and other types of heterofermentative lactobacilli. It was proposed, therefore, to refer to them as a new type, *Lac. kefir.* The number of mesophilic lactobacilli in kefir starters does not exceed 10^2-10^3/ml, and these micro-organisms do not play an important role in the formation of starters or product quality (Koroleva, 1975, 1988). Of the obligate homofermentative types, *Lac. bulgaricus* and *Lac. helveticus* (syn. *Thermobacterium*) are found in kefir starters in quantities up to 10^4-10^5/ml. These micro-organisms grow in number at elevated temperatures of fermentation (Koroleva, 1975, 1988).

(c) Mesophilic, heterofermentative, lactic acid streptococci (*Leuconostoc mesenteroides* and *Leu. mesenteroides* sub-sp. *dextranicum*) participate in the formation of the specific taste and aroma of kefir, and they may, on superfluous development, cause gas formation. *Leu. mesenteroides* sub-sp. *dextranicum* is activated at elevated temperatures and by high yeast counts. Strains of *Leu. mesenteroides* producing polysaccharide were isolated by Rosi and Rossi (1978). Using immunologic and DNA–DNA studies, Hontenbeyrie and Gasser (1975, 1977) showed that *Leu. mesenteroides* is different from all the leuconostocs, except *Leu. dextranicum*. Taking this into consideration, the strains belonging to this type should be determined as subspecies of *Leu. mesenteroides* and called *Leu. mesenteroides* sub-sp. *dextranicum*.

(d) Yeasts (*Kluyveromyces marxianus* sub-sp. *marxianus, Torulaspora delbrueckii, Saccharomyces cerevisiae, Candida kefir*) have been found and identified by a number of authors (Kaminski, 1955; Khrul'kevich & Khrul'kevich, 1959; La Riviére, 1963; Rosi, 1978a). Differences in the species detected are explained by the variety of cultivation methods employed. Yeasts take an active part in maintaining symbiosis among the micro-organisms, CO_2 formation in kefir (Koroleva, 1966; Gobbetti *et al.*, 1986) and development of the specific taste and aroma. Their excess growth may cause gas formation in the product and packaging problems.

(e) Acetic acid bacteria (*Acetobacter*) are very active in maintaining symbiosis between the kefir starter microfloras. Rosi (1978b) found that *A. aceti* is the only species present in kefir. Schulz (1946) studied the effect of these micro-organisms on the growth of the kefir grains, and the maintenance of starter activity. According to Bavina and Rozhkova (1973), acetic acid bacteria improve the consistency of the kefir by increasing its viscosity. In case of excessive growth of the acetic acid bacteria, evident viscosity and slime may appear in the kefir starter and/or kefir.

Basic difficulties in studying and monitoring the kefir starter microflora lie in the fact that its components, separated as pure cultures, do not grow in milk (acetic acid bacteria, non-lactose fermenting yeasts), or decrease sharply their biochemical activity (heterofermentative lactic acid streptococci, lactobacilli). Therefore, the effect of separate groups of kefir starter micro-organisms on product quality and chemical composition has to be studied by artificially increasing or decreasing this or that group, and changing the cultivation conditions. The relationship between the major groups may be summarized as follows (Koroleva, 1975):

— lactic acid bacteria are stimulated by yeasts and acetic acid bacteria at the expense of protein peptonization and vitamin development—also partial lactic acid utilization;
— non-lactose fermenting yeasts and acetic acid bacteria can grow in milk only whan lactic acid bacteria are present hydrolysing lactose; lactic acid accumulation creates more favourable conditions for their development;
— homofermentative lactic acid streptococci inhibit, to a certain extent, yeast formation and make for slow alcohol accumulation.

Basic Changes in Kefir Resulting from the Activity of its Microflora and Creating Therapeutic Properties

Lactose Fermentation
Lactic acid development, as a result of the activity of the lactic acid microflora, makes all fermented products more favourable for people suffering from genetically stipulated lactose intolerance. Kefir possesses this property, like other fermented milk products, and in kefir, the lactic acid microflora is dominated by mesophilic, homofermentative, lactic streptococci. Thus in kefir, unlike yoghurt, L(+) lactic acid is formed, and depending on the amount of lactic acid, kefir can have either a purging or constipating effect.

CO_2 Formation
At kefir production, the growth of yeasts and heterofermentative lactic streptococci results in CO_2 development, and this promotes 'fine flake' coagulum formation which, in turn, stimulates protein digestion (Davidov & Sokolovsky, 1968) and appetite.

Alcohol Formation
Alcohol in kefir is a result of yeast activity in cases where the lactic streptococci are inhibited by the production process (e.g. high acidity), or by using tightly capped bottles and agitation. Such conditions for kefir production were created at small dairies at the beginning of the century, and the alcohol content in kefir was 1–2%. The technology used at present in the USSR eliminates the possibility of alcohol accumulation in noticeable quantities; its content in kefir is 0·01–0·05%.

Free Amino Acid Accumulation
During the first hours of fermentation, the free acids of milk are inten-

sively consumed, but at further ripening, their accumulation is reported (Ilyenko-Petrovskaya, 1965). This is explained by the fact that, at the initial stage of fermentation, homofermentative lactic streptococci require free amino-acids for growth, but as ripening goes on, development of the proteolytically active microflora—yeasts and acetic acid bacteria—causes an accumulation of free amino acids and other milk protein proteolysis products.

Vitamin Formation
While in fermented products manufactured with pure cultures of lactic bacteria the vitamin content is lowered as a rule, in kefir, accumulation of B and P (riboflavenoid) group vitamins occurs due to the activity of yeasts and acetic acid bacteria (Koroleva, 1975).

Antibiotically Active Agents
According to Bukanova (1955), yeasts (*Torulaspora* type) separated from kefir possessed pronounced antibiotic activity against coli bacteria. Ivanova (1975) showed that, of all the kefir starter microflora components, the mesophilic, homofermentative, lactic streptococci (*Str. lactis* and *Str. lactis* sub-sp. *cremoris*) and the acetic acid bacteria were the most antibiotically active against coli bacteria; the antagonism had both bacteriostatic and bactericidal characters. Lactose fermenting yeasts also inhibited the growth of coli bacteria. Non-lactose fermenting yeasts did not have this property. Antagonism of the kefir starter microflora showed itself to be more pronounced than the antagonism of its separate components.

Polysaccharide Formation
During recent years, there has been growing interest in microbial polysaccharides and their possible application as gelling agents in the food industry. Of all the microbial polysaccharides, dextran, xanthan and aubasidan can be used in foods (Elinov, 1985). Polysaccharides are found to have a favourable effect on humans by binding water and some toxic agents, and also by lowering the cholesterol level in blood (Elinov, 1985). La Riviére et al. (1967) separated a polysaccharide from kefir grains called 'kefiran'. It consisted of glucose and galactose in equal proportion, dissolved slowly in cold water and quickly in hot water; a 2% solution is a viscous liquid. Kefiran is present in the capsule material of the big, rod-like bacteria that are predominant in kefir grains, and which have the

characteristics of *Lactobacillus*. Toba *et al.* (1986) obtained identical polysaccharides from kefir grains and two strains of lactobacilli—*Lac. kefir* and *Lac. kefir* K_1 separated from kefir grains. Fudziya (1981) obtained a sticky polysaccharide sediment from kefir grains by organic solvent extraction, and used it to improve the consistency of kefir.

Thus, the most probable producers of polysaccharides in the kefir grain are representatives of *Lactobacillus*. The possibility of polysaccharide formation in the starter and kefir as a result of the activity of *Leuconostoc* (dextran, xanthan) and *Acetobacter aceti* (glucan) cannot be excluded either.

Polysaccharides are found in kefir grains up to 34% (Vorobjeva *et al.*, 1987) and in kefir, 0·2–0·7% (Kononovich *et al.*, 1986). The effect of polysaccharides obtained from kefir grains on proteolytic activity in the gastro-intestinal tracts of rats has been studied (Strelkova *et al.*, 1987). Two hours after injection (12 mg per 1 kg animal mass) of the polysaccharide, protease activity in gastric juice and the small intestine increased more than twice as compared with the controls. In tests *in vitro*, the polysaccharide raised trypsin esterase activity by 35–55%.

Research done by Shiomi *et al.* (1982) on mice showed that the soluble polysaccharide separated from kefir grains, when taken orally, inhibited Erlich sarcoma and sarcoma 180 (S-180) development as compared with controls by 40–59% and 21–81%, respectively. Tumour growth was inhibited also after intra-peritoneal injection. Mirotushi *et al.* (1986) stated that the oral intake of polysaccharide separated from kefir grains raised immunity in mice and suppressed malignant tumour development. Studies on kefir enrichment with polysaccharides have been undertaken, but a major obstacle in using kefir grain polysaccharides for the improvement of fermented products is their poor solubility. Kononovich *et al.* (1986) suggested that kefir should be enriched by polysaccharides from kefir grains by producing large amounts of them in milk. In the All-Union Research Institute for Dairy Industry, a kefir grain treatment was developed that made it possible to get an homogeneous mass of grains/ polysaccharide well mixed with the fermented milk coagulum.

Regimes for Starter and Kefir Production

The complicated and symbiotic microbiological composition of kefir grains explains why it is so difficult to get the starter with the constant and optimal composition necessary for standard kefir manufacture. It is quite evident that any deviation from the established regimes of kefir grain cultivation will lead to changes in the microbiological composition of the starter and, consequently, to changes in the character and duration of milk fermentation and product quality.

In the USSR, comprehensive studies have been undertaken in order to establish how cultivation regimes, e.g. cultivation temperatures, grains-to-milk ratio, duration and conditions for keeping a starter plus grains, number of agitations of milk plus grains to be applied in the course of a fermentation, washes of kefir grains, etc., will influence the kefir starter microflora (Koroleva, 1988b).

When a rather large quantity of kefir grains is inoculated (1:20; 1:10), the fermentation process becomes shorter but, in this case, the level of homo- and heterofermentative lactic streptococci and yeasts is lower than at 1:50; this can be explained by the fast accumulation of lactic acid and the short process duration. Thermophilic lactobacilli and the acetic acid bacteria, do not, as a rule, respond to changes in the starter:milk ratio.

When the quantities of kefir grains inoculated in milk decrease, there has been noted a tendency for the major groups of micro-organism to increase. The most active development of all the groups takes place at a grain-to-milk ratio of 1:50. The highest enrichment of kefir starter with volatile fatty acids and CO_2 takes place when the grains-to-milk ratio is 1:30 and, especially, 1:50. This fact can probably be connected with the more intensive growth in the kefir starter of yeasts and heterofermentative lactic streptococci. At elevated cultivation temperatures, growth of the homofermentative lactic acid (streptococci and lactobacilli) bacteria becomes more intense. Values of pH rapidly decrease to the level inhibitory to homofermentative and heterofermentative lactic streptococci, which is why their numbers in the finished kefir starter are lower. For other, more acid-resistant groups, elevated temperatures are much more favourable.

To accumulate all the principal groups in the starter, it is quite enough to hold the starter, together with the grains, at the fermentation temperature for 5–6 h after coagulum formation. A further 24-h incubation of the starter with the grains at comparatively low temperatures (7–10°C) does not, on the whole, change the microflora in either character or quantitative composition.

An additional agitation during the course of cultivation of the grains in milk increased the mesophilic, homofermentative lactic streptococci and yeasts in the starter by ten times. The additional agitation did not affect the heterofermentative lactic streptococci, thermophilic lactobacilli and acetic acid bacteria. Stirring prevents extraneous mould growth on the starter surface, and promotes a more even distribution of the metabolic products of the microflora.

Weekly washing of the grains with potable water led to a sharp decrease in the major starter groups. It also decreased starter activity, made longer the fermentation process and worsened the taste and consistency of the end product. The normal starter microflora was restored only 3–5 days later. Yeasts were the only micro-organisms which did not diminish after washing; sometimes even their stimulation was observed.

This study shows that it is possible to directly regulate the microflora of a kefir starter by changing the conditions of cultivation. At present, in the USSR method of kefir manufacture that provides a high quality product with typical taste, aroma and consistency (Koroleva, 1988a), 1 ml of kefir starter must contain: homofermentative, mesophilic, lactic streptococci (10^8–10^9); thermophilic lactobacilli (10^5); heterofermentative lactic streptococci (10^7–10^8); yeasts (10^5–10^6); acetic acid bacteria (10^5–10^6). The starter acidity must be in the range of 105–110°.

The main features of the production of a starter with the above-mentioned microflora should be as follows (Koroleva et al., 1971; Koroleva, 1988b):

— milk must be renewed every day at the same time;
— the grains-to-milk ratio should be kept in the range of 1:30–1:50;
— the skim-milk used for cultivation must be pasteurized at 95°C for 10–15 min;
— cultivation temperature 18–20°C;
— agitate 2/3 times in the course of the cultivation process.

A kefir starter prepared by the cultivation of kefir grains in milk is called 'grainy' (Starter 1). As a rule, Starter 1 is used for kefir production. Sometimes a bulk starter (Starter 2) is used which is prepared by fermenting pasteurized milk with Starter (1). At elevated temperatures (25–27°C), the milk fermentation takes 6–8 h, the mesophilic, heterofermentative lactic streptococci and yeasts do not have time enough to develop, and the kefir acquires an atypical taste.

At 20–22°C, the fermentation takes 10–12 h. If the coagulum is cooled slowly for about 10–12 h, an accumulation of heterofermentative lactic streptococci and yeasts occurs, and the kefir acquires the pronounced taste and aroma. If the coagulum is cooled quickly, these micro-organisms do not develop, and the taste again becomes atypical. Thus, the total cycle of kefir production should take some 20–24 h, and include a fermentation at a temperature not exceeding 25°C for 10–12 h, and ripening with slow cooling for 10–12 h.

The maximum quantity of homofermentative, mesophilic lactic streptococci has been detected in kefir made with milk inoculated with 1 or 2% of starter. It can probably be explained by the fact that these smaller quantities of starter create more favourable conditions for growth of this group of micro-organisms. Similar results are obtained when heterofermentative lactic streptococci are studied. The levels of other micro-organisms, especially those of the thermophilic lactobacilli and acetic bacteria in kefir were the same, and did not depend on the quantity of Starter 1 inoculated.

When Starter 1 was used for kefir production, the content of homofermentative lactic streptococci and yeasts was higher than in analogous samples of kefir made with Starter 2. The quantity of acetic acid bacteria was the same, but numbers of thermophilic lactobacilli were higher in the samples of product prepared with Starter 2. In so far as the thermophilic lactobacilli lead to excess acid accumulation in kefir, their increase cannot be considered to be a positive factor.

These results explain why it is reasonable to use Starter 1 for kefir production and, at the same time, decrease its quantity. At present, Starter 1 is mainly used for kefir production.

Milk for kefir manufacture is selected according to microbiological, organoleptic, physical and chemical properties. It is pasteurized at 85–87°C for 5–10 min or at 90–95°C for 2–3 min. Homogenization takes place at 12·5–17·5 MPa. The milk is then cooled to the temperature of 22–25°C, and inoculated with Starter 1 (by 1–3%) or Starter 2 (by 3–5%).

The milk is fermented for 8–12 h until the coagulum acidity reaches 90–100°; then after agitation, slow cooling begins, which lasts for 10–12 h.

One millilitre of kefir, manufactured by existing technology, contains: mesophilic, homofermentative, lactic streptococci (10^9); thermophilic lactobacilli (10^5); heterofermentative lactic streptococci (10^7–10^8); yeasts (10–10^5) and acetic acid bacteria (10^4–10^5).

Kefir is produced with fat contents of 3·2; 2·5; 1·0 and < 0·1%, and the acidity of the finished product is usually not more than 90–100°. Kefir has a homogeneous consistency, and specific refreshing taste. In kefir manufactured by existing technology, the alcohol content does not exceed 0·01–0·05%, and the CO_2 content is comparatively low. However, these data can be easily modified by corresponding changes in starter microflora, fermentation and ripening temperature, and method of bottle-capping.

Therapeutic Properties of Kefir

Thanks to the presence of homofermentative, lactic streptococci, yeasts and acetic acid bacteria in kefir, it possesses high antibiotic activity against extraneous micro-organisms in the intestine. The last two groups of micro-organisms cause increased proteolysis, and the accumulation of free amino acids and other products of protein hydrolysis. They also form vitamins of the B and P groups. Development of the specific kefir microflora (yeasts and heterofermentative lactic streptococci) results in CO_2 formation, which results in a 'fine flake', dispersed coagulum structure, that improves digestion of the product; the presence of CO_2 and alcohol also stimulates the appetite.

Kefir consumption facilitates diuresis, decreases the specific gravity of urine, raises the excretion of urea and other nitrogen metabolism products, as well as chlorides and phosphates, and increases urine excretion (Davidov & Sokolovsky, 1968; Samsonov & Budagovskaya, 1982). The microbial polysaccharides present in kefir facilitate the binding of toxic agents, and decrease the cholesterol content in blood (Elinov, 1985). The kefir microflora produces mainly L(+) lactic acid, which is physiologically acceptable to humans.

The slightly acidic taste of kefir and its characteristic microflora improve salivation, the excretion of stomach and pancreatic enzymes, and peristalsis. Kefir contributes to a more even movement of food in the intestine, and the lactic acid and antibiotic substances inhibit 'rotting processes' in the small intestine (Davidov & Sokolovsky, 1968). Kefir with low acidity (up to 90°) can act as a purging agent, but with higher acidity (110–120°) as a constipating agent. At present, kefir with an acidity level of 90–100° is produced in the USSR, and this product has a normalizing effect on the intestines.

According to publications in 'Minity', 'Emiury' and some other newspapers in Japan, Kubo has studied the effect of freeze-dried kefir on cancer development in mice. It was found that, in the case of kefir administered at the rate of 100 mg per kg of weight, the efficiency of cancer inhibition was 52·8–53·6%, e.g. the data obtained were higher than the standard evaluation of 50% established by the International Cancer Research Association. Parallel administration of kefir and an anti-cancer medicine, mithomicin, weakened collateral effects and showed higher efficiency, some 67%, as compared to mithomicine taken alone (58%). Kefir consumption also stimulated microphages and improved immunity. It had a favourable effect on the resistance of humans to catching colds and other diseases.

In the USSR, kefir is widely used in hospitals and sanatoria. It is included in diets of patients suffering from intestinal diseases, metabolic disorders, namely diabetes, hypertension, atherosclerosis and allergic diseases. The product is traditionally used in the diet of children from an early age.

KUMYS

The History of Kumys Production
Kumys, prepared from mare's milk by a combined fermentation (lactic acid and alcohol), is an ancient drink widespread in Eastern Europe and Asia. The Scythian tribes which roamed from place to place in South-East Europe and Middle Asia about 25 centuries ago used to drink mare's milk in the form of a drink called 'kumys'. The famous Greek historian Gerodot wrote, in the 5th century BC, that 'kumys is a favourite Scyths' drink'. They made their prisoners whip mare's milk in high wooden vessels, and it was used as a nutritive and weakly alcoholic drink.

In the Ipatiev chronicle (1182), it was written that Prince Severskyi was taken prisoner, but the soldiers who guarded him drank kumys, got drunk and fell asleep. The Prince managed to escape. In 1269, Marco Polo mentioned kumys to be a 'pleasant milk drink'.

For many centuries, roaming tribes prepared kumys in tursuks—leather sacks of 25–30 litres capacity made from the thigh part of a horse hide, i.e. having a broad bottom and narrow, long sleeve. In the Caucasus, these sacks are called 'burduks'. Freshly-drawn mare's milk was filled into the tursuk, and agitated with a paddle inserted into it; as the tursuk was emptied, it was refilled with fresh milk. Emptying and refilling of the tursuks was done many times until the kumys acquired a bad taste. Then the tursuk was emptied completely, washed and smoked over a fire of birch bark, dry leaves and grass. Settled Bashkirs used wooden barrels (chelyaks) made of lime or oak wood, in which the kumys was agitated by wooden paddles.

At the end of the last century, kumys production was started in Russia at specialized hospitals built in the regions of traditional horsebreeding. At present, kumys from mare's milk is prepared in sanatoria for tuberculosis patients; the milk for this purpose coming from specialized farms or from their own stud farms.

Much kumys from mare's milk is prepared in Mongolia, where about 0·5 million milking mares are available. Kumys is an important national product, and annually some 13·5 million litres are produced; some 30% of the raw milk resources being used (Baldorzh & Namsray, 1980).

Due to the limited availability of mare's milk and seasonal fluctuations in its production, intensive research work has been done in the past decade to develop formulae, methods and starters for kumys preparation from cow's milk. At present, it is produced on a limited scale at some dairies.

Starter

For a long time, kumys production in the regions of developed horse-breeding—at home and in hospitals adopting kumys treatments—used natural starters, i.e. inoculation of the raw milk with an existing product.

In 1940–1960s, after the pasteurization of mare's milk had been developed, a starter composed of pure cultures of *Lac. bulgaricus* and yeasts was used for kumys manufacture (Voitkevich, 1948).

Studying the microflora of kumys, Krisanfova (1961) found that kumys yeasts consisted of three main types: lactose fermenting—*Saccharomyces lactis*; non-lactose fermenting—*Sac. cartilaginosus*; non-carbohydrate fermenting—*Mycoderma*. The major micro-organisms participating in the lactic acid process were thermophilic lactobacilli of the *Lac. bulgaricus* type. It was also stated that mesophilic lactobacilli close to *Lac. causasicum* were present. Krisanfova (1965) recommended that *Sac. lactis, Lac. bulgaricus* and *Lac. causasicum* should be included in the starter for cow's milk kumys.

In Mongolia, lactic streptococci, *Str. lactis* type, and lactobacilli, *Lac. bulgaricus* type, are present in starter cultures for kumys production from mare's milk. However, Voitkevich (1948) and Bogdanov (1962) found the use of lactic streptococci in kumys starters unreasonable. The high acidity of mare's milk immediately after fermentation and its low buffering power create unfavourable conditions for the development of streptococci. It was also found that some strains of lactic streptococci inhibit yeast growth and hinder alcohol production. Bannikova and Lapshina (1970) developed a starter for cow's milk kumys manufacture at dairies. The starter consists of the yeast, *Sac. lactis*—antibiotically active against *Mycobacterium tuberculosis, Lac. bulgaricus* and *Lac. acidophilus*. The latter are introduced into the starter to increase the antibiotic activity of kumys against pathogens of the gastro-intestinal tract.

Effect of Milk Composition and Starter Microflora on the Therapeutic Properties of Kumys

As compared with cow's milk, mare's milk has a quite different composition. It has lower casein and fat contents, but higher levels of albumin, lactose, vitamins C, B_1, B_2, B_{12}, and P, and micro-elements (cobalt and copper). Casein is present in kumys in the form of fine flakes, which are not felt by the tongue. As a result, the product consistency is liquid, and the nitrogenous substances it contains are easily digested by humans. Mare's milk differs from cow's milk in its 'immunity' to *Mycobacterium tuberculosis*; it is known that horses rarely contract tuberculosis. The development of the lactic acid bacteria leads to the accumulation of lactic acid and antibiotically-active agents. Yeasts produce CO_2 and alcohol; *Sac. lactis* is most active in alcohol formation (2–3·5%). *Sac. lactis* and *Sac. cartilaginosus* are capable of producing antibiotically active agents.

Kumys from Cow's Milk

The composition of cow's milk for kumys production is made similar to mare's milk by lowering the fat and casein contents, and increasing the whey protein content (Selesnev & Artykova, 1967; Schamgin *et al.*, 1978). Definite compensation for the lack of 'immunity' against tuberculosis is achieved by starter selections having high antibiotic activity against *Mycobacterium tuberculosis*—yeasts (*Sac. lactis* type) and *Lac. acidophilus* (Koroleva, 1988*b*).

In domestic conditions (Kazakhstan, Bashkiria, Mongolia), the starter for mare's milk kumys is prepared in the following way (Skorodumova, 1961). At the beginning of the season (in Spring) small tursuks containing goat's milk kumys left from the previous season and kept in a cold place, are half-filled with mare's milk; the next day, the other half is added. Three days later, the starter is usually restored, and later on, kumys of high quality is used as a starter.

Pure cultures for mare's milk kumys are prepared by adding yeasts and lactic acid bacteria to freshly pasteurized milk cooled to 31–35°C. The milk is agitated for 10–12 min, capped with cotton wool, and kept at 25–28°C with agitation at intervals of 2–3 h; 24 h later, the starter is ready. Its acidity is 120–130°, and the alcohol content 1–2% (Skorodumova, 1961). In Mongolia, the recommended value for starter acidity for mare's milk kumys production is 110–115° (Baldorzh & Namsray, 1980). The starter for cow's milk kumys is prepared at dairies under conditions that provide even growth of all the micro-organisms composing it. To this end, pasteurized milk (300 ml) cooled to 30°C, is inoculated with 10–15 ml of pure cultures of *Lac. acidophilus, Lac, bulgaricus* and yeasts. The milk is agitated and kept at 30°C for 7–10 h, then at room temperature for 3–6 h. The starter obtained is used for bulk starter making by inoculating 10–20% of it into skim-milk pasteurized at 85–90°C and cooled to 30°C. After the acidity of the milk reaches 85–90°, the starter is agitated and held for 3–4 h for yeast growth. During the course of this time, the starter is stirred at least ten times and for 5 min on each occasion. The acidity of final starter is in the range of 110–140°.

Mare's milk kumys is produced in the following way: a kumys starter prepared with pure cultures (after ripening and cooling to 6–8°C and with an acidity of about 140°) is inoculated into fresh or pasteurized mare's milk (temperature 31–35°C), so that the acidity of the inoculated milk is brought to 45–50° and the temperature to 25–26°C. After starter addition, the milk is agitated for 15–20 min. Then, in the course of the first hour of fermentation, it is agitated 3–4 times for 1–2 min until the young product acquires its specific taste. The kumys is then filled into bottles, capped with corks, and left for alcohol fermentation and gas accumulation to take place. Finally, the filled bottles are placed in a cooling chamber where the lactic acid process stops, and only a weak alcohol fermentation continues.

Mare's milk kumys has a liquid/gaseous consistency, and a specific, 'biting', refreshing taste; the acidity of the finished product is about 100–140°, alcohol content 0·2–2·5%.

Kumys produced in Mongolia has the following, average characteristics: acidity, 112·6°; density, 1·006; fat content, 2·0%; total protein, 1·9%; lactose, 3·2%; alcohol, 1·6%; casein, 1·4%; vitamin C, 77·5 mg/kg (Baldorzh & Namsray 1980).

Cow's milk kumys is prepared from skim-milk (according to Method 1) (Seleznev & Artykova, 1967), and a special milk mixture for kumys (according to Method 2) (Schamgin *et al.*, 1978).

Method 1

Kumys is produced from skim-milk with 2·5% sucrose added before pasteurization. After pasteurization at 90–92°C for 2–3 min and cooling to 28–30°C, the mixture is inoculated with 10% of starter with constant stirring for 15–20 min. The milk is fermented at 26–28°C for 5–6 h until coagulum formation with an acidity of 75–85°. The coagulum is agitated, aerated and cooled. For this purpose, ice water is supplied between the walls of the tank, and the coagulum is mixed by stirring with simultaneous air saturation. The first agitation is stopped 10–15 min after a homogeneous coagulum consistency has been achieved.

The coagulum is then agitated every 15–20 min, without aeration, for 1·5–2 h. At the end of the process, the temperature is lowered to 16–18°C, and the acidity is 85–95°. The product acquires a liquid, homogeneous, slightly foaming consistency. Kumys is filled into glass bottles, with a narrow neck, of 0·5 litres capacity and capped with a crown cork. The temperature of ripening and storing does not exceed 4°C. Kumys produced by Method 1 is characterized by a milky-white colour, pure lactic acid taste, slight aroma of yeast, and no extraneous flavours. Weak kumys may have a slightly sweet taste. After agitation, the consistency of the product is homogeneous with fine protein particles (which are not felt by the tongue), and it is a little foamy. The acidity of the product is 100–120° (weak), 120–140° (average), 140–150° (strong); the alcohol content is not limited.

Method 2
According to Method 2, cow's milk kumys is prepared from a mixture of liquid components including whole milk, skim-milk and cheese whey. To facilitate handling, the mixture is dried. Before drying, it contains (dry matter basis): whole milk, 34·6%; skim-milk, 0·8%; non-separated cheese whey, 64·6%; the fat content should be 0·9–1%, and the density, 1·022–1·023 kg/m^3. Before using, the mixture is dissolved in potable water warmed to 50–55°C. It is then pasteurized at 85–87°C with 5–10 min holding, homogenized at 10–12 MPa and cooled to 32–34°C. Inoculation and fermentation take place in tanks; 20% of starter and ascorbic acid (0·2 g/kg dissolved in water) are added to the mixture. The mixture is fermented at 28–30°C and constantly agitated for 3–4 h until the acidity reaches 75–80°. After the fermentation stops, alcohol formation is stimulated by cooling the kumys to 16–18°C with ice-cold water circulation through the vessel jacket, and holding at this temperature with constant agitation for 1–2 h. The end product is filled into 0·5-litre glass bottles and capped with foil. Ripening takes place at 6–8°C. It has a milky-white colour, a clean, refreshing lactic taste, but with a slight taste and aroma of yeast. Product consistency is homogeneous with fine protein particles and slight effervescence. The fat content of the finished product is 1·5% and the solids-non-fat content 9%. The acidity of weak kumys is 95°, and the alcohol content should be not less than 0·6%; medium-acidity, 110°, alcohol content, 1·1%; strong acidity, 130° acidity and 1·6% alcohol content.

By chemical composition, the product is close to mare's milk kumys in fat, whey proteins, casein and vitamin C content, but non-hermetic closure does not allow the product to obtain the foaming consistency characteristic of mare's milk kumys.

Therapeutic Properties of Kumys
The anti-tuberculosis properties of mare's milk kumys were known long ago, and were confirmed by the experience of many generations of Russian practitioners. The beginning of kumys preparation in Russia dates back to 1858, when the first sanatorium with kumys treatment was founded by N. V. Postnikov near the town of Samara.

Kumys therapy improves both the secretion of gastric juices and gastro-intestinal peristalsis and has a favourable 'tonic' effect. Kumys proteins, being peptonized and finely dispersed, are easily digested and assimilated by humans. There is several times more vitamin C in mare's milk kumys than in other fermented products. Physicians came to the conclusion that normalization of digestive tract activity during courses of kumys treatment of tuberculosis patients was partially the result of diminishing, and even eliminating, the tuberculosis intoxication that accompanies lung tuberculosis, and often causes stomach diseases.

Kumys therapy facilitates normalization of the white corpuscle level in blood, and the erythrocyte sedimentation rate; as a rule, the haemoglobin content in blood is raised, and peripheral nerve system function is also improved.

Kumys treatment has a positive effect on tuberculosis of the upper respiratory tract, and there is a tendency for the infection to be halted. It also has a beneficial therapeutic effect in cases of gastro-intestinal disease not connected with lung tuberculosis, and is good for chronic bronchitis, pneumonia, dry pleurisy and for recuperation after other infections.

Clinical tests with kumys prepared from cow's milk by Method 2 showed that its therapeutic effect is similar to that of mare's milk kumys; it is recommended as a supplementary remedy for tuberculosis and the treatment of several other diseases.

ACIDOPHILIN

The technology of acidophilin production was developed at the Moscow dairy named after Gorky M. When acidophilin production and starter composition were studied, the necessity of lowering the fermentation temperature as compared to acidophilus milk production (37°C) was taken into consideration, with a view to getting a low acidity product (Koroleva, 1966).

Milk for acidophilin production, after pasteurization and cooling to 32–35°C, is inoculated with a starter composed of three types in equal proportions: *Lac. acidophilus*, active in milk fermentation, *Str. lactis* and a kefir starter; the total amount of the inoculated starter is 2–5%. The fermentation continues until a dense coagulum is formed, and an acidity of 75–80° is achieved. Fermentation takes 6–8 h. The product is filled into bottles or packages and cooled to 8°C; the final product acidity is 100–110°. Rods and streptococci are found in microscopic preparations in approximately equal quantities; the yeast content of acidophilin does not exceed a few hundred, so with microscopic analysis, they may not be detected, and can be found only by growing on special media.

Acidophilin is successfully used, along with *acidophilus* milk, for the treatment of colitis, enterocolitis, both of staphylococcal and other origin, dysentery, disbacteriosis and other intestinal diseases.

Acidophilus-Yeast Milk
The starter for *acidophilus*-yeast milk and the technology of its production were developed in 1952 by Skorodumova (1961), and it was then adjusted for the conditions of the industry (Bogdanova & Bogdanova, 1982). As a starter, strains of *Lac. acidophilus* active in coagulating milk, have high antibiotic activity against undesirable intestinal microfloras and give rise to a milk of viscous consistency, are used; the yeast, *Sac. lactis*, antibiotically active to *Mycobact. tuberculosis*, is also included.

The primary starter is prepared by the inoculation of sterilized milk with 5–10% of viscous cultures of *Lac. acidophilus* and yeast washed from the surface of an agar slant. The milk is fermented at 25–30°C for 25 h; 3–5% of the resultant starter is put into sterilized milk to obtain a secondary starter. Control of the starter includes examination of microscopic preparations, which should contain large numbers of rods and 5–6 yeast cells/field. If the yeast level in the starter decreases, it is necessary to extend its ripening at 25–30°C. When yeast development is too intense, the period of ripening should be shortened. Milk for *acidophilus*-yeast milk production is pasteurized and cooled to 33–35°C, and then inoculated with 3–5% of starter and agitated carefully; the fermentation lasts for 4–6 h at 30–32°C until a coagulum with 70–80° acidity is formed. The fermented milk is cooled to 10–17°C, with agitation, and left to ripen for 6–10 h. The acidity of the finished product should be 90–100°. The examination of microscopic preparations should show only rods and yeasts.

Acidophilus-yeast milk has a viscous consistency, and a slightly acid, sharp, yeasty taste. The highest antibiotic activity was discovered in 3-day-old products.

Acidophilus-yeast milk is recommended as an additional remedy for tuberculosis, intestinal diseases and furunculosis.

REFERENCES

Baldorzh, R. & Namsray, P. (1980). *Mongolian kumys*. Ulan-Bator, State Publishing House, p. 115.
Bannikova, L. A. & Lapshina, L. V. (1970). *Works of VNIMI*, **27**, 78–82.
Bavina, N. & Rozhkova, I. (1973). *Molochnaya Promyshlennost*, **2**, 15.
Bogdanov, V. M. (1962). *Microbiology of Milk and Milk products*. Pischepromizdat, Moscow.
Bogdanova, E. A. & Bogdanova, G.I. (1982). *Fermented Products Manufacturing*. Pychevaya Promyshlennost, Moscow.
Bukanova, V. P. (1955). *Gig. Sanit.*, **8**, 32–6.
Davidov, R. B. & Sokolovsky, V. P. (1968). *Milk and Milk Products in Human Nutrition*. Medicina, Moscow, 236.
Elinov, N. P. (1985). In *Proceedings of Symposium on Studies in Microorganisms Biomass for Nutrition*. Puschino, pp. 60–72.
Fudziya, K. K. (1981). Authorship Certificate No. 56-29503, Japan.
Gobbetti, M., Rossi, J. & Clementi, F. (1986). Proceedings of the XXI International Dairy Congress, The Hague, p. 182.
Hontenbeyrie, M. & Gasser, F. (1975). *Int. J. Syst. Bacteriol.*, **25**(1), 1–6.
Hontenbeyrie, M. & Gasser, F. (1977). *Int. J. Syst. Bacteriol*, **27**(1), 9–14.
Ilyenko-Petrovskaya, T. P. (1965). *Applied Biochemistry and Microbiology*, **1**(2), 246–7.
Ivanova, L. N. (1975). *Moloch. Prom.*, **9**, 8–11.
Kaminski, G. (1955). *Przen. Spoz.*, **9**(7), 258–64.
Kandler, O. & Kunath, P. (1983). *Syst. Appl. Microbiol.*, **4**(2), 288–94.
Krisanfova, L. P. (1961). *Konevodstvo*, 317.
Krisanfova, L. P. (1965). *Molochnaya Promyshlennost*, **3**, 38.
Khrul'kevich, V. & Khrul'kevich, A. (1959). *Moloch. Prom.*, **20**(12), 24–5.
Kononovich, N. G., Kamaljan, M. G., Dmitrovskaya, G. P., Kosenko, L. B. & Zakharova, I. J. (1986). *Molochnaya Promyshlennost*, **12**, 24–5.
Koroleva, N. S. (1966). *Technical Microbiology of Fermented Milk Products*. Pischepromizdat, Moscow.
Koroleva, N. S. (1975). *Technical Microbiology of Dairy Products*. Pischevaya Promyshlennost, Moscow.
Koroleva, N. S. (1988a). Starters for fermented milk. Sec. 4. Kefir and kumys starters. Bulletin of IDF N 227, 35–40.

Koroleva, N. S. (1988b). Sec. VII Technology of kefir and kumys. Bulletin of IDF N 227, pp. 96–100.
Koroleva, N. S., Bavina, N. A. & Milutina, L. A. (1971). Authorship Certificate No. 314–380. Patented in Great Britain (Patent No. 1300315).
La Riviére, J. W. M. (1963). *J. Gen. Microbiol.*, **31**(1).
La Riviére, J. W. M., Kooiman, P. & Schmidt, K. (1967). *Arch. Microbiol.*, **59**(5), 269–78.
Mirotushi, M., Mixuguchi, I., Aibara, K. & Matuhasi, T. (1986). *Immunopharmacology*, **12**, 29–35.
Rosi, J. (1978a). *J. Sci. Tec. Latt. Casear*, **29**, 59–67.
Rosi, J. (1978b). *J. Sci. Tec. Latt. Casear*, **29**, 221–7.
Rosi, J. & Rossi, J. (1978). *J. Sci. Tec. Latt. Casear*, **29**, 291–305.
Samsonov, M. & Budagovskaya, V. (1982). *The Role of Fermented Dairy Drinks as Dietary Products*. XXI Int. Dairy Congr. Moscow, Vol. 2, pp. 140–5.
Schamgin, V. K., Zalashko, L. S., Mochalova, K. V., Pastukhova, L. M., Abramovskaya, A. K., Rozhkova, I. A., Vojtovich, G. A. & Antipova, S. Y. (1978). Brief communication at the XX Intern. Dairy Congr. Paris, p. 266.
Schulz, M. (1946). *Milchwissenschaft*, **1/2**(1), 19–27.
Seleznev, V. I. & Artykova, L. A. (1967). *Molochnaya Promyshlennost*, **12**, 36–9.
Shiomi, M., Sasaki, K., Murofushi, M. & Aibara, K. (1982). *Jap. J. Med. Aci. Biol.*, **35**, 75–80.
Skorodumova, A. M. (1961). *Dietetic and therapeutic fermented milk products*. Medizdat, Leningrad.
Strelkova, M. A., Vorobjeva, A. I., Komov, V. P. & Vitovskaya, G. A. (1987). Studies in structure and biological activity of galactoglucans from kefir grains. All-Union Conference: Carbohydrates chemistry and Biochemistry, Puschino, p. 236.
Toba, T., Arihara, K. & Adachi, S. (1986). Proceedings of the XXI International Dairy Congress, The Hague, p. 183.
Voitkevich, A. F. (1948). *Microbiology of Milk and Milk Products*. Pischepromizdat, Moscow, p. 320.
Vorobjeva, A. I., Vitovskaya, T. A., Elinov, N. P., Korshunov, A. I., Filchakova, S. A. & Koroleva, N. S. (1987). *Molochnaya Promyshlennost*, **7**, 14–16.

INDEX

Acetic acid bacteria, 163
Acetobacter, 24
Acetobacter aceti, 33
Acidophilin, 176–8
Acidophilus Milk, 25
Acidophilus products, 81–116
  categorization of, 83–6
  factors affecting survival, implantation and maintenance of acidophilus microflora in humans, 93–100
  health promoting properties, 100–11
  milk based, 84
  other than milk based, 85
  possible adverse effects, 114
  survival and implantation in humans, 91–3
  survival of *Lactobacillus acidophilus* in, 87–91
  therapeutic, 86
  see also *Lactobacillus acidophilus*
Acidophilus-yeast milk, 177
Alcohol formation, 164
Amino acids, 4, 6
Antibiotics, 165
  side effects of, 58, 128
Anti-carcinogenic factors, 103–5
Antimicrobial factors, 102–3
Auto-toxication theory, 45–6
Avian infectious bronchitis virus, 105

Baby foods, 145–6
Bacillus, 30
Bacillus bifidus, 117
Bacillus bifidus communis, 117
Bifidobacteria, 50, 51, 53, 117–57
  activity in the intestines, 152
  anti-tumour activity of, 133–5
  beneficial competition with other intestinal bacteria, 129
  beneficial functions in the intestine, 124–6
  dietary adjuncts, 126
  differentiating characteristics of, 118
  enterobacteria, and, 127
  fermented milk products containing, 146–51
  health properties associated with, 123–37
  human species, 118–23
  improved N retention and weight gain in infants due to, 126
  influence of carrier food on viability, 144
  influence on intestinal microflora, 129–30
  inhibition of nitrate reduction, 126
  intestine in adults, in, 129–37
  intestine of infants and small children, in, 123
  occurrence and species distribution, 121–3
  possible adverse effects, 137
  potentially beneficial roles, 177
  potentially therapeutic effects, 128–9
  prophylactic and therapeutic aspects, 130–3
  protective effects against enteric infections, 127
  relevant species, 119–21
  resistance tests, 139
  selection of strains as dietary adjuncts, 137–9
  survival of intestinal passage, 139
  therapeutic aspects of fermented milks containing, 117
  to prevent side-effects of antibiotic therapy, 128
  transit in gastro-intestinal tract, 141–4

Bifidobacteria-containing products, 144–51
Bifidobacterium, 24, 31–2
Bifidobacterium adolescentis, 32
Bifidobacterium bifidum, 28, 32, 39, 57, 58, 131, 133, 145–51
Bifidobacterium breve, 32, 130
Bifidobacterium dentium, 137
Bifidobacterium infantis, 32
Bifidobacterium longum, 32, 131, 133, 136, 137, 146–8, 150
Bifidus-Acidophilus-Mesophilic Streptococcus-Milk, 151
Bifidus-Acidophilus-Milk, 148–9
Bifidus-Acidophilus-Pediococcus Milk, 150
Bifidus-Acidophilus-Thermophilus-Milk, 149
Bifidus-Acidophilus-Yoghurt, 150
Bifidus milk, 146–8
Bifidus-Thermophilus-Milk, 149
Bifidus-Yoghurt, 150
Bifighurt, 149
Bifiline, 146
Bile acids, 54–5
Bilkys, 150
Biocines, fermentation, 50–1
Biogarde, 149

Calcium, 12, 14–15
Campylobacter, 19
Cancer therapy, 58
Candida, 131, 132
Candida albicans, 57
Candida kefyr, 33
Candida utilus, 135
Carcinogens, 59
β-carotene, 11
β-casein, 4
k-casein, 4, 12
Caseins, 3
β-casomorphins, 4
Cholesterol, 60, 76–7, 110–11
Citrobacter spp., 131
CO_2 formation, 164
Colon cancer, 58–9, 106

Colon tumourigenesis, 135
Colonization of lumen content, 142
Constipation, 59–60
Cow's milk, 1, 2
  kumys from, 173–6
  principal proteins in, 3
  principal vitamins in, 16
Crohn's disease, 53
Cultura, 35, 148

Denaturation, 18

Ehrlich Ascites tumours, 78
Escherichia coli, 53, 58, 129, 137
Essential amino acids, 7, 8

Fatty acids, 9–11
Fermentation, 45, 47–51
  changes resulting from, 68–70
  micro-organisms used in, 47
  nutritional and physiological effects, 51
Fermented milk products, 159
  containing bifidobacteria, 146–51
  home made, 46
  industrially-made, 46
  types of milk used, 46
Fermented milks, 46
  history of, 23
  micro-organisms of, 23–43
Fibronectin, 6
Free amino acid accumulation, 164–5

β-galactosidase, 75
Gastric acidity, 54
Gastro-intestinal organisms, 91
Growth promotion, 100–1

Health food-microbial adjuncts, 83
Health hazards, 19
Health promoting properties, 100–11
Heat treatments, 17

INDEX

Heavy metals, 14
Human flora, 51–3
  establishment of, 52–3
  protective functions, 53
Human milk, 2

Immune system, 136
Immunoglobulins, 6
Immunological response, 105–10
Infantile diarrhoea, 57
Infectious canine hepatitis, 105
Inoculation of process milk, 34–9
Intestinal flora
  composition of, 54
  re-establishment of, 56–7
Intestinal microflora, influence of yoghurt starter bacteria, 72–3

Kanamycin, 131
Kefir, 160-71
  basic changes creating therapeutic properties, 164–6
  grains, 31, 161, 166
  production, 160, 166–9
  starter, 161–4, 166–9
  therapeutic properties of, 170
Klebsiella spp., 131
Kluyveromyces, 33
Kumys, 171–6
  effect of milk composition and starter microflora on therapeutic properties of, 173
  from cow's milk, 173–6
  production, 171–2, 174–6
  starter, 172
  therapeutic properties of, 176

α-lactalbumin, 5
Lactana-B, 145
Lactase, 12, 49
Lactic acid bacteria, 47, 69, 91, 102, 159
Lactic acid fermentation, 49, 65

Lactobacilli, 53–62
  effects of administration, 56
  prevention and therapy, 56–62
Lactobacillus, 24–9
Lactobacillus acidophilus, 25, 28, 39, 50, 51, 57–61, 66, 83, 131, 149, 150, 151, 173, 174, 177
  factors affecting survival, implantation and maintenance of acidophilus microflora in humans, 93–100
  survival and implantation in humans, 91–3
  survival in products, 87–91
  therapeutic activity of, 111–13
  see also Acidophilus products
Lactobacillus brevis, 28, 50, 162
Lactobacillus bulgaricus, 25, 29, 49, 51, 65, 66, 68–73, 78, 89, 162, 172, 174
Lactobacillus casei, 28, 59, 78
Lactobacillus causasicum, 172
Lactobacillus delbrueckii sub-sp. *bulgaricus*, 25, 159
Lactobacillus helveticus, 28, 135, 162
Lactobacillus kefir, 28, 162, 166
Lactobacillus plantarum, 50
Lactococcus lactis, 30
Lactoferrin, 5–7
β-lactoglobulin, 5, 6
Lactolose, 126
Lactose, 11–12, 49, 60–1, 73–5, 101–2
Lactulose, 12
Leuconostoc, 24, 30–1
Leuconostoc mesenteroides, 31, 163
Linolenic acid, 11
Lipids, 60
Listeria, 19
Liver tumourigenesis, 133
Lumen content, colonization of, 142
Lysine, 18
Lysozyme, 5

Maillard reaction, 18
Measles, 105
Micelle structure, 4

Micro-organisms of fermented milks, 23–43
Milk
  as food, 1–22
  constituents of, 2
  history of, 1
  nutritional value, 70–1
  trace elements in, 13–14
Milk carbohydrates, 11–13
Milk fat degradation, fermentation, 48
Milk lipids, 8–11
Milk minerals, 13–14
  nutritional role of, 14–15
Milk proteins, 2–7
  fermentation, 48
  nutritional value of, 7–8
Milk solids, consumption, 20
Mineral absorption, 102
Mycobacterium tuberculosis, 173, 177
Mycoderma, 172

Net protein utilisation (NPU), 7
Newcastle disease, 105
Non-protein nitrogen (NPN), 6
Nutritional value, effect of processing, 17–18

Pasteurisation, 17, 18
Pediococcus, 24, 32
Pediococcus acidilacti, 32
Pediococcus acidilactici, 150
Peptides, 4
pH values, 51–2
Phage-inhibitory media, 35
Phage-resistant media, 35
Pharmaceutical preparations, 151
Polysaccharide formation, 165–6
Progurt, 151
Propionibacterium, 24, 32
Protein efficiency ratio (PER), 7
Proteus spp., 131
Pseudomonas spp., 131

Riboflavin, 3–4

Saccharomyces cartilaginosus, 172, 173
Saccharomyces lactis, 172, 173
Salmonella, 19
Salmonellosis, 19
Staphylococcus auerus, 30
Starter cultures, 33–4, 36, 65
  categories of, 47
  see also under specific cultures
Steatorrhoea, 49
Streptococcus, 24, 29–30
Streptococcus lactis, 29. 30, 151
Streptococcus thermophilus, 25, 29, 39, 49, 65, 66, 68–73, 89, 149, 159
Stress effects, 62
Sweet Acidophilus Milk, 25, 89

Technical starters, 47
Therapeutic starters, 47
Toxic amines, 59
Trace elements in milk, 13–14
Transgalactosylated oligosaccharide (TOS), 130
Traveller's diarrhoea, 57
Triacylglycerides, 9
Trichomoniasis vaginalis, 150
Tuberculosis, 19
Tumour inhibition by yoghurt starter bacteria, 77–9
Typhoid fever, 19

UHT processes, 18
Ultra-high temperature (UHT) processes, 17

Vesicular stomatitis, 105
Virus inhibition, 104–5
Vitamins, 11, 15–17
  fermentation, 50
  formation, 165
  yoghurt manufacture, in, 69–70
Volatile acids, 55
Volatile fatty acids, fermentation, 49

Weaning causes, 51
Whey proteins, 4–5

Xanthine oxidase, 6

Yeasts, 33, 159, 163, 174, 177
Yersinia, 19
Yoghurt, 45, 65–80
  anti-microbial actions produced by starters, 71–2
  anti-tumour activity of starter bacteria, 77–9
  characterization, 65
  containing bifidobacteria, 150
  definition, 65
  effect on lactose utilization by humans, 74
  flavoured, 75
  frozen, 75
  hypocholesterolaemic effect of, 76–7
  manufacture of, 66–8
  nutritional value, 70–1
  role of starter cultures, 72
  see also Starter cultures